Establishing a Sovereign Guided Weapons Enterprise for Australia

International and Domestic Lessons Learned

CHRISTOPHER A. MOUTON, CARL RHODES, MARK V. ARENA,
PAUL DeLUCA, ANDREW DOWSE, JOHN P. GODGES, ADAM R. GRISSOM,
CALEB LUCAS, ERIK SILFVERSTEN

Prepared for the Australian Department of Defence

For more information on this publication, visit **www.rand.org/t/RRA1710-1**.

About RAND Australia

The RAND Corporation (Australia) Pty Ltd is RAND's subsidiary that does work for Australian clients on defence, national security, health, education, sustainability, growth, and development. To learn more about RAND Australia, visit www.rand.org/australia.

Research Integrity

Our mission to help improve policy and decisionmaking through research and analysis is enabled through our core values of quality and objectivity and our unwavering commitment to the highest level of integrity and ethical behaviour. To help ensure our research and analysis are rigorous, objective, and nonpartisan, we subject our research publications to a robust and exacting quality-assurance process; avoid both the appearance and reality of financial and other conflicts of interest through staff training, project screening, and a policy of mandatory disclosure; and pursue transparency in our research engagements through our commitment to the open publication of our research findings and recommendations, disclosure of the source of funding of published research, and policies to ensure intellectual independence. For more information, visit www.rand.org/about/principles.

RAND's publications do not necessarily reflect the opinions of its research clients and sponsors.

Library of Congress Cataloging-in-Publication Data is available for this publication.
ISBN: 978-1-9774-0829-7

Cover: Flag - railwayfx/Adobe Stock, Javeline missile – army.mil/U.S. Army, F/A-18 - U.S. Navy National Museum of Naval Aviation/U.S. Navy, NASAMS - Lance Cpl. Ujian Gosun/U.S. Marine Corps, STARStreak - Sgt Mark Webster RLC/Defence Imagery, SM-2 - U.S. Navy/Wikimedia Commons

About This Report

In response to Australia's 2020 Defence Strategic Update calling for increased weapon inventories across the Australian Defence Force, this project offers an initial examination of key considerations related to the creation of a Sovereign Guided Weapons Enterprise. This exploration consists of three focus areas: (1) lessons learned from five international sovereign defence industry initiatives; (2) lessons learned from Australia's own experiences developing sovereign shipbuilding and munitions manufacturing industries, particularly the steps needed to implement acquisition plans cost-effectively and the resulting economic and industry effects throughout Australia; and (3) the implications of all these lessons for the Australian Sovereign Guided Weapons Enterprise, particularly the challenges it is likely to face and the key considerations for creating a plan to address those challenges.

This main report covers the first and third focus areas. Specifically, this report describes the relevance of the five international case studies to Australia. These case studies—of comparable enterprises in Japan, the United Arab Emirates, the United Kingdom, Canada, and Norway—review the most relevant aspects of each country's experience and its lessons for Australia. This report concludes with a synthesis of the overall lessons, both international and domestic, for an Australian Sovereign Guided Weapons Enterprise.

Two limited-distribution appendices provide further details on the relevance of the two domestic case studies to an Australian Sovereign Guided Weapons Enterprise. Appendix A focuses on the Australian Naval Shipbuilding Enterprise, while Appendix B focuses on Australia's munitions manufacturing industry.

This research was conducted by RAND Australia in conjunction with the RAND National Security Research Division (NSRD), and sponsored by the Australian Department of Defence.

About RAND Australia

RAND Australia is RAND's Canberra-based subsidiary that analyses defence, national security, economic, and social issues for Australian clients. With a commitment to core values of quality and objectivity, RAND Australia combines local research talent with world-class experts from across RAND's global presence to solve complex Australian public policy problems.

For more information on RAND Australia or to contact our director, please visit www.rand.org/australia.

About the RAND National Security Research Division

This research was undertaken in the Acquisition and Technology Policy Center (ATP) of RAND's NSRD, which conducts research and analysis for the U.S. Office of the Secretary of Defense, the U.S. Intelligence Community, the U.S. State Department, allied foreign governments, and foundations.

For more information on RAND NSRD, see www.rand.org/nsrd.

Contents

Figures and Tables

Figures

Tables

Summary

In response to Australia's 2020 Defence Strategic Update calling for increased weapon inventories across the Australian Defence Force (ADF), and the Australian government announcement of a Sovereign Guided Weapons Enterprise, RAND Corporation explored key considerations related to the creation of this enterprise. This exploration focused on three areas: (1) lessons learned from five international sovereign defence industry initiatives; (2) lessons learned from Australia's own experiences developing sovereign shipbuilding and munitions manufacturing industries; and (3) the implications of all these lessons, international and domestic, for the creation of an Australian Sovereign Guided Weapons Enterprise.

Maintaining control over the production and sustainment of essential defence capabilities is an existential concern for any state. Contemporary scholarship in the fields of defence economics and defence innovation emphasises the singular nature of national defence acquisition systems. Every such system is unique, because each is subject to distinctive strategic, political, social, cultural and economic dynamics at the national, institutional and programmatic levels. No two nations confront the exact same strategic problems, nor do they possess precisely the same resources or assets. National institutions, processes and norms are the products of different histories and particular circumstances. As a result, each national defence acquisition system is bespoke to the nation it serves. Each looks and works a bit differently. Care must therefore be taken in attempting to apply insights or 'lessons' across national defence acquisition systems. That is, in fact, a major component of the first key lesson learned.

Overall, we draw seven lessons for establishing a Sovereign Guided Weapons Enterprise. First and foremost, we recognise the need for such an enterprise to be bespoke to the Australian domestic and strategic context. We also recognise the complexity of creating such an enterprise. The remaining lessons focus heavily on ensuring sustainable economic conditions for a sovereign enterprise, as well as prioritising partnerships and collaborations.

Lesson 1: Define desired outcomes as part of developing a bespoke sovereign enterprise. In developing a sovereign enterprise, Australia first needs to define and prioritise its desired outcomes. Australia's strategic problems, situated at the fulcrum of the emerging Pacific Century, are unique in the international system. Australia's defence-related politics, its systems and institutions for defence acquisition, and its technological strengths and weaknesses are all distinct from those of even its closest allies. Sovereign defence acquisition will therefore mean something different to Australia than it does to other nations.

Lesson 2: Sovereignty needs to be carefully defined. Operational sovereignty, in the case of guided weapons, is about removing the risks of losing access to, or control over, needed capabilities. While Australia might need access to, or control over, the requisite elements required for producing certain weapons, the capabilities do not necessarily need to be produced in Australia or by Australian companies to achieve operational sovereignty. However, there may be other benefits (jobs, assured supply chains, etc.) that may come with an enterprise of building weapons in Australia.

Lesson 3: Complex sovereign enterprises take time and effort to build. Expertise to develop world-class weapons can take decades to develop, particularly when establishing a sustainable export business. Development is reliant upon supporting innovation and education systems; yet even with sufficient expertise, investments in complex weapon systems can take a decade or more to come to fruition. There is also the ongoing need to govern all parties involved in a complex sovereign enterprise, including myriad types of organisations inside and outside of government.

Lesson 4: Affordability and sovereignty need to be balanced. A sovereign enterprise will likely come with additional costs. If Australia's domestic defence industry develops and manufactures a relatively small number of guided weapons for the ADF, the weapons will likely have high unit procurement and sustainment costs. While this approach would enhance domestic control and supply chain security, it would likely incur an above-market price premium.

Lesson 5: Joint development offers advantages and limitations. Joint development with allies and partner nations and their industries is attractive for two reasons. First, having common systems can aid interoperability and mutual logistic support, reducing localised wartime supply shortfalls. Second, each partner of a joint development team can learn from the others—if information is shared freely. On the other hand, these partnerships are often stymied by concerns about proprietary information being unduly withheld or, conversely, used inappropriately. Moreover, joint programs do not promote product differentiation, which has both strategic and economic negative consequences.

Lesson 6: Offsets can spur growth under certain circumstances. A continually adapting offsets program can enable defence-sector growth and encourage joint ventures that lead to the transfer of critical technology. Offset programs have been used to spur the development and growth of domestic defence companies. Offsets have also facilitated joint ventures with international partners, and these ventures have enabled critical technology transfers, including intellectual property related to precision-guided munitions. Despite these notable successes, offset policies can impose additional costs and may not always achieve their aims.

Lesson 7: Industrial capacity needs to be right-sized. As Australia considers a Sovereign Guided Weapons Enterprise, its capacity should be consistent with domestic requirements. The alternative would be an orphaned production capacity that would unnecessarily increase unit costs and act as a deadweight within the guided weapon acquisition budget. Production costs could be reduced through export activities, but such activities would need to be permitted under technology controls, and there would need to be a likely market. Thus, right-sizing will require deliberate analysis, forecasts and decisions to produce a capability that is sufficient to meet Australian needs and adaptable to compete in the international marketplace, but not too big as to increase fixed costs that could not be recouped.

Acknowledgements

We are extremely grateful for the support we received from CDRE Nigel Smith and COL Ruth Perry. Their clear framing of the research question was instrumental to our effort, and we are extremely appreciative of the thoughtful insights and perspectives they provided throughout this project.

This effort was truly a team effort, and we are grateful for the entire RAND National Security Research Division (NSRD) and RAND Australia team. We are thankful for Joel Predd's constructive oversight of this project, which helped us address some key challenges. We are especially thankful for Yun Kang's support, ensuring an effective and efficient quality assurance process on a compressed timeline. None of this would have been possible without the fantastic work of the NSRD operations team, specifically Michelle Platt, Norma Ockershausen and Nancy Pollock. The RAND Australia operations team, Josh Crawford and Amy Grogan, were also instrumental to the effort.

Finally, we would like to thank Michael Kennedy and Roger Lough for their detailed and thorough review of our work, as well as Chad Ohlandt for inputs he provided early in this project. While any errors or omissions are solely the responsibility of the authors, the critiques from these reviewers helped sharpen our research and highlighted key areas for further exploration.

CHAPTER ONE

An Australian Sovereign Guided Weapons Enterprise

In response to Australia's 2020 Defence Strategic Update calling for increased weapon inventories across the Australian Defence Force (ADF), this project offers an initial examination of key considerations related to the creation of a Sovereign Guided Weapons Enterprise. This examination consists of three focus areas: (1) lessons learned from five international sovereign defence industry initiatives; (2) lessons learned from Australia's own experiences developing sovereign shipbuilding and munitions manufacturing industries, particularly the steps needed to implement acquisition plans cost-effectively and the resulting economic and industry effects throughout Australia; and (3) the implications of all these lessons for the Australian Sovereign Guided Weapons Enterprise, specifically the challenges it is likely to face and a tentative plan to address them. This main report covers the first and third focus areas, while two limited-distribution appendices cover the second focus area.

The examination provided in this main report is not meant to enumerate a list of definitive recommendations based on similar experiences in other countries but rather to highlight areas for consideration and broad early insights (or 'overall lessons') for the Sovereign Guided Weapons Enterprise. To support ongoing enterprise deliberations, we also craft a framework for the enterprise, based on established best practices.

Background

On 31 March 2021, the Morrison government announced the creation of a Sovereign Guided Weapons Enterprise with an initial investment of AUD 1 billion as part of a broader AUD 270-billion, ten-year defence investment. The Prime Minister highlighted two key benefits of this enterprise: first, the creation of new skilled jobs and, second, the securing of Australia's defence capabilities. Minister for Defence Peter Dutton described the ultimate objective of the enterprise as ensuring 'we have adequate supply of weapon stock holdings to sustain combat operations if global supply chains are disrupted' as well as increasing and improving ADF operational capacity. The need for resiliency against supply chain disruptions was made acute by the COVID-19 pandemic, which, as the Prime Minister described, showed that 'having the ability for self-reliance, be it vaccine development or the defence of Australia, is vital to meeting our own requirements in a changing global environment'.[1]

Less than five months later, Minister for Defence Industry Melissa Price published an op-ed piece on 10 August 2021 echoing the above priorities by describing the 'multifaceted benefits of the sovereign guided missiles project',[2] also known as the Sovereign Guided Weapons Enterprise. According to Minister Price, the key benefit and primary purpose of the enterprise lies in 'making sure the Australian Defence Force is trained, equipped and ready to mobilise' for 'whatever the world might throw at us' in a 'challenging strategic defence environment'. This will necessitate achieving 'domestic self-reliance' in the age of COVID-19, 'as supply chains are disrupted around the world' and developing 'an industry that can manufacture components and guided weapons here at home'. Minister Price argues that the enterprise will generate additional benefits for Australia by creating 'demand for hi-tech, local manufacturing jobs' and 'opportunities for Australian businesses across a range of sectors', while also providing 'opportunities to supply key components to our key international strategic partners', thereby generating 'significant export revenue to Australia'.

The Sovereign Guided Weapons Enterprise initiative builds on the 2020 Defence Strategic Update, which called for the Department of Defence (Defence) to improve the resilience of its enterprise to shocks and outside interference, including in the supply and inventory of weapons:

> To further build resilience and self-reliance, Defence will increase the range and quantity of the weapon stocks it holds. Funding has also been allocated for exploring and potentially implementing additional measures, including the development of sovereign manufacturing capabilities for advanced guided weapons and explosive ordnance and expanding ADF fuel storage capacity.

Likewise, the 2020 Force Structure Plan directed the Defence to take specific steps in support of the goals defined in the Strategic Update:

- Increase weapon inventory across the ADF to ensure weapons stock holdings are adequate to sustain combat operations if global supply chains are at risk or disrupted.
- Redevelop the Mulwala explosives and propellant facility to expand its capacity for munitions production, thus enhancing the resilience of ADF ammunition supply.
- Explore the potential for a new sovereign guided weapons and explosive ordnance production capability to mitigate supply risks, especially for those munitions with long lead times.

---

[1] Prime Minister, Minister for Defence, Minister for Defence Industry, Minister for Industry and Minister for Industry Science and Technology, 'Sovereign Guided Weapons Manufacturing', press release, 31 March 2021.

[2] Melissa Price, 'The Multifaceted Benefits of the Sovereign Guided Missiles Project', *Defence Connect*, 10 August 2021.

A foundational principle of the Morrison government's vision for a Sovereign Guided Weapons Enterprise is that Australia 'will work closely with the United States on this important initiative to ensure that we understand how our enterprise can best support both Australia's needs and the growing needs of our most important military partner'.[3] This vision was cemented during the 31st Australia–United States Ministerial Consultations (AUSMIN) Consultations, when Defence Minister Dutton announced that the United States and Australia had agreed to cooperate on the development of Australia's guided weapons and explosive ordnance enterprise. This cooperation is part of a broader cooperation across the key pillars of the alliance, namely 'science, technology, strategic capabilities, and defence industrial base integration'.[4]

There is no doubt that the Australian government is motivated to establish a Sovereign Guided Weapons Enterprise. To make it happen will require an alignment of that motivation with the nation's defence budgets, the capabilities of the nation's overall defence industrial base, and the nation's policies regarding sovereignty and related matters. The bulk of this report, therefore, will take stock of how these factors have aligned or have failed to align, first in a series of five international case studies and then in a pair of domestic Australian enterprises.

But a foundational question first needs to be asked: What exactly does *sovereignty* mean in the context of an Australian guided weapons enterprise? With that question addressed, the remainder of this introduction will present a 'sovereign weapon development framework' for Australia, a recognition of the unique Australian context (and the unique context for every sovereign weapons program), a description of the methodology employed for this analysis, and a roadmap for the chapters to follow. In framing our analysis, we note that, to date, a specific delineation of 'guided weapons' has not been provided; therefore, we take a broad view of what qualifies as guided weapons. Specifically, we do not explicitly limit guided weapons to only surface-to-air, surface-to-surface and air-to-ground precision weapons, but include potentially loitering munitions and underwater weapons as well.[5]

Sovereignty in Context

Sovereignty has many connotations but is used primarily in relation to Westphalian sovereignty. That is, it concerns a state's ability to exercise power exclusive of interference by external sources.[6] Sovereignty is a term used regularly in defence publications and is mentioned throughout the 2018 Defence Industrial Capability Plan and the 2020 Defence Strategic

---

[3] Prime Minister, 2021.

[4] Antony J. Blinken, Lloyd Austin, Marise Payne and Peter Dutton, 'Joint Press Availability', press release, Dean Acheson Auditorium, Washington, D.C., 16 September 2021.

[5] This broad view offers the potential for guided weapons to exist across a wide spectrum of speed, range and warhead sizes, to include small special operations precision-guided munitions (SOPGMs) and much larger and more sophisticated hypersonic weapons.

[6] Stephen Krasner, 'Rethinking the Sovereign State Model', *Review of International Studies*, Vol. 27, 2001, p. 20.

Update. This concept of sovereignty was explored by Dowse et al., and a revised version of their work appears in this section.[7]

The *2018 Defence Industrial Capability Plan* defines sovereignty in this fashion:

> Sovereignty is about the independent ability to employ Defence capability or force when and where required to produce the desired military effect. Australia seeks the ability to maintain, employ, sustain and upgrade our Defence capabilities with the maximum level of Australian access to, or control over, the essential skills, technology, intellectual property, financial resources and infrastructure so that the Australian Defence Force is positioned to achieve the Strategic Defence Objectives.[8]

The second sentence is important: While Australia might need 'access to' or 'control over' the 'skills, technology', etc., associated with certain weapons, the capabilities do not necessarily need to be produced in Australia or by Australian companies. Rather, sovereignty is about removing the *risks* of losing access to, or control over, those needed capabilities. Such risks to availability may be mitigated either through domestic industrial production or through other options, such as stockpiling supplies, using alternative means to achieve an effect, increasing supplies in direct response to indications and warnings, or reaching agreements that increase trust and responsiveness in supply arrangements. For some capabilities, the cost and viability of a domestic industry approach to sovereignty may pose constraints; in such cases, the other options for supply assurance may be preferred. This assured support to the military has been referred to as *operational sovereignty*.[9] While the authors are not aware of any work to date to quantify this risk, operations research methodologies could aid in quantifying these risks and hence allow for an optimised set of solutions to minimise those risks.

Neither the Defence Industrial Capability Plan's emphasis on 'independent' operations nor the references in Defence white papers to 'self-reliance' imply that defence capabilities must be produced by Australian industry in order to preserve sovereignty. However, their recognition of the important contribution of local industry to achieving sovereignty is a continuation of Defence policies over at least the past quarter of a century.[10]

Asserting that certain strategic industry capabilities needed to remain resident in Australia, the 2009 Defence White Paper introduced the concept of Priority Industry Capabilities, or PICs. PICs were defined as 'those industry capabilities that would confer an essential

---

[7] Andrew Dowse, Tony Marceddo and Ian Martinus, 'Cyber Security and Sovereignty', *Australian Journal of Defence and Strategic Studies*, Vol. 3, No. 2, 2021.

[8] Defence, *2018 Defence Industrial Capability Plan*, 2018, p. 17.

[9] Graeme Dunk, 'The Decline of Trust in Australian Defence Industry', *Australian Defence Magazine*, 10 February 2020.

[10] Discussion of requirements and policies associated with self-reliance have featured in Defence strategic reviews and white papers from the late twentieth century; see Paul Dibb, 'The Self-Reliant Defence of Australia: The History of an Idea', in Ron Huisken and Meredith Thatcher, eds., *History as Policy: Framing the Debate on the Future of Australia's Defence Policy*, Canberra, ANU Press, 2007.

strategic capability advantage by being resident within Australia, and which, if not available, would significantly undermine defence self-reliance and ADF operational capability'.[11]

The PICs policy met with initial concern about the lack of detailed process and the omission of capabilities that were not high-profile but were nonetheless strategically important.[12] A 2015 parliamentary inquiry found that the PICs were confusing and passive, that there was a gap between policy and action, and that specific areas of interest were outdated.[13] Among 11 recommendations aimed at improving defence industry policy, the parliamentary committee recommended the PIC program be discontinued.

In 2016, Defence announced the Sovereign Industry Capability Assessment Framework (SICAF) as the means by which sovereign industrial capabilities supporting the ADF would be identified and developed.[14] Immediate concerns with the SICAF included its narrower definition of capabilities compared with the PICs, as well as its perceived lack of innovative focus in reviewing sovereign capabilities against emerging strategic risks.[15]

The 2018 Defence Industrial Capability Plan provided greater transparency to the SICAF and its ten priorities, describing it as a logical process comprising strategic guidance, assessment of capability programs against sovereignty criteria, prioritisation, and identification of resultant priorities. Support and assistance programs would be developed for each priority based on a specific implementation plan, with assistance available through the Centre for Defence Industry Capability, the Defence Innovation Hub and the Next Generation Technologies Fund.[16]

The Defence Industrial Capability Plan describes each of the ten Sovereign Industrial Capability Priorities (SICPs) in a single paragraph, with an intent to release implementation and industry plans for each priority. Eight of the ten plans had been published by April 2021, five of these being released in the last two months of 2020.[17]

The six criteria used in SICAF to assess capability programs are protection of intent,[18] independence of action, interoperability, assurance of supply, skills retention, and competitive

---

[11] Defence, 'Defence White Paper 2009', 2009, p. 128.

[12] Leigh Purnell and Mark Thomson, *How Much Information Is Enough? The Disclosure of Defence Capability Planning Information*, Australian Strategic Policy Institute, December 2009, pp. 63–64.

[13] Joint Standing Committee on Foreign Affairs, Defence and Trade, *Principles and Practice—Australian Defence Industry and Exports: Inquiry of the Defence Sub-Committee*, Parliament of the Commonwealth of Australia, 2015, pp. 25, 40, 42 and 43.

[14] Defence, *2016 Defence Industry Policy Statement*, p. 23.

[15] Graeme Dunk, 'Defence Industry Policy 2016—Well-Intentioned but Conflicted', *Security Challenges*, Vol. 12, No. 1, 2016, pp. 145–146.

[16] Defence, 2018, pp. 31–46.

[17] Defence, 'Implementation and Industry Plans', webpage, 2020c. There was an expectation that all ten plans would be released by the end of 2020; see Centre for Defence Industry Capability, 'Land Combat and Protected Vehicles Sovereign Industrial Capability Priority Plan Released', webpage, 3 May 2021.

[18] *Protection of intent* refers to the ability to employ military capability without divulging military intent (Defence, 2018, p. 31).

advantage. Once that process identifies potential industry capabilities, they are prioritised against a range of filters, including whether they help deter, deny and defeat threats; the importance of Australian control; whether they sustain current capability; and lead-time implications.

These criteria and filters focus on sovereignty and security, but such capabilities delivered by a domestic defence industry can also foster broader benefits, as well as economic and knowledge spillover effects.[19] While the focus remains on the core Sovereign Industry Capability, as Figure 1.1 illustrates, there are important peripheral benefits for the broader Australian Industry Capability (AIC).

Minister Price's characterisation of the Sovereign Guided Weapon Enterprise as delivering both security and economic benefits combines these concepts of operational sovereignty and Australian industry. Importantly, however, and in contrast with Figure 1.1, the potential exists for supply chain arrangements to confer access and control and to contribute to operational sovereignty without being delivered through a domestic industry.

FIGURE 1.1

Sovereign Industry Capability and Australian Industry Capability

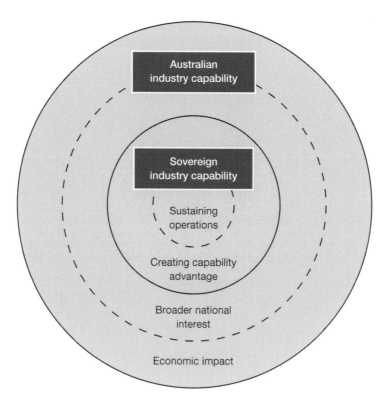

SOURCE: Adapted from Government of South Australia, 2020, p. 5.

---

[19] Joint Standing Committee on Foreign Affairs, Defence and Trade, p. 12.

There are two additional considerations for a sovereign industrial capability: time and trust.

First, current thinking is that drivers of industry capability may evolve slowly over time, while strategic challenges may evolve rapidly. The 2020 Defence Strategic Update acknowledged that previous assumptions of a ten-year strategic warning of conflict were no longer valid, heralding a new era in which conflict may arise at short notice and involve periods of high-intensity warfare.[20] The implications include the need for protection against short-notice threats, plus the capacity to rapidly ramp up delivery of systems to mitigate the effects of attrition. These short-notice threats demand a greater emphasis on force design, stock holdings and replenishment than on formal acquisition processes.

Second, Australia's current ability to trust and verify systems is limited, particularly given the fragility of international supply routes and, ironically, the sophistication of modern information systems within almost every ADF capability. These supply chain risks include cyber-security risks, which have been characterised as the ADF being taken out of any meaningful fight before even getting to it.[21] While an AIC to produce all weapon systems domestically (a task that is likely not achievable) could mitigate many of these concerns, it is prudent to develop trust mechanisms to mitigate the risks associated with reliance on international supply. Such mechanisms would align with the SICAF priority of exerting sovereign control to assure *availability* of these capabilities, vice an assumption of *manufacturing* all the supporting technologies domestically.

A Sovereign Weapon Development Framework for Australia

Our framework for the development of sovereign weapons in Australia details the actions, partners and infrastructure needed to effectuate an Australian precision-guided missile (PGM) program. Our approach is adapted from a capability development framework designed by RAND Australia for the Department of Home Affairs[22] and aligned with the One Defence Capability Model (ODCM).[23] We modify these generalised processes to create a bespoke approach for sovereign weapons and PGMs specifically.

The approach by Dortmans et al. focuses on how to achieve a given capability, which they define as 'the capacity and intent to achieve and sustain a desired effect or output to meet one or more strategic objectives'. They attribute this definition to the Department of Home Affairs. Their framework, which builds on Australia's existing policies, involves two

---

[20] Defence, *2020 Defence Strategic Update*, 2020a, p. 14.

[21] Marcus Thompson, 'Information Warfare—a New Age?', speech delivered at the Military Communications and Informations Systems Conference, Canberra, Australia, 15 November 2018.

[22] Peter Dortmans, Jennifer D. P. Moroney, Kate Cameron, Roger Lough, Emma Disley, Laurinda L. Rohn, Lucy Strang and Jonathan P. Wong, *Designing a Capability Development Framework for Home Affairs*, Santa Monica, Calif.: RAND Corporation, RR-2954-AUS, 2019.

[23] Defence, *Defence Capability Manual*, 2020b.

tasks that emanate from government guidance: how to identify capabilities and how to design them, thereby effectively addressing the strategic need identified by the government.

Dortmans et al. developed this framework through a detailed examination of the current Defence approach, lessons from international capability life-cycle management models, reviews of scholarly literature, case studies from other contexts, and interviews with Australian officials and relevant individuals in the United Kingdom and the United States. The resulting framework integrates three key elements of capability development: fundamental inputs to capability (FIC), key enablers, and assurance.

We incorporated the aspects of this framework that are applicable to developing a sovereign guided missile program and aligned it with the development stages identified in the *Defence Capability Manual*, particularly the ODCM. Our adapted framework largely includes the same stages of the development process as those in the ODCM model, progressing from (1) strategy/concepts to (2) risk mitigation/requirement setting to (3) acquisition to (4) in-service/disposal. The notable difference between the ODCM and our framework is the addition of program strategy as a stage in the process. We present program strategy as occurring early in the research and development (R&D) phase. This stage necessarily needs to occur early to formulate approaches to tasks such as securing investments, creating new technology, developing the capacity to manufacture at scale, and storing the munitions.

To capture issues salient to the enterprise, our framework widens the focus on capability design, to explicitly include both R&D and 'manufacturing & sustainment' along with 'training & education', as shown in Figure 1.2. We position training and education such that it undergirds the entirety of the process's second stage—that is, training in sustainment feeds back to future capability management and moves the process forward again. This positive feedback loop is a critical component of effectuating a sustainable sovereign PGM program. Moreover, we explicitly integrate continued efforts at sustaining the enterprise into the stage at which the product is in-service. Consistent with the work of RAND researchers Cook et al., assurance and contestability are features across all stages in our framework.[24] Figure 1.2 shows how the actions (or processes) on top straddle the required partners and FIC infrastructure components, all of which are grounded in assurance and contestability.

Finally, as itemised in the lower tiers of Figure 1.2, we detail the support elements—the FICs and key enablers—that help generate the sovereign PGM capability. We do this by explicitly listing partners and infrastructure necessary to stand up a sovereign PGM program. The FICs include facilities and operational support (infrastructure) required for a capability. We outline four components of infrastructure that are critical to a sovereign PGM program: production facilities; test and evaluation facilities; storage and distribution sites; and disposal sites. Each of these FICs contributes to a critical component of the larger effort. For the key enablers, we identify a range of partners with cross-cutting functions that contribute to the overall enterprise. These key enablers include Defence, the Australian munitions industry,

---

[24] Cynthia R. Cook, Emma Westerman, Megan McKernan, Badreddine Ahtchi, Gordon T. Lee, Jenny Oberholtzer, Douglas Shontz and Jerry M. Sollinger, *Contestability Frameworks: An International Horizon Scan*, Santa Monica, Calif.: RAND Corporation, RR-1372-AUS, 2016.

FIGURE 1.2

Sovereign Weapon Development Framework

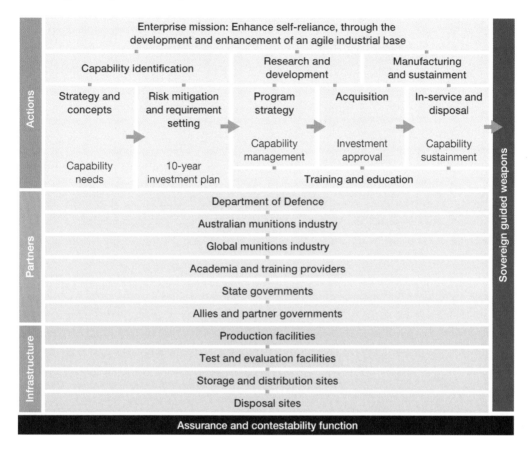

the global munitions industry, academia and training providers, Australian state governments, and allies and partner governments. Taken together, this framework maps the processes that could effectuate an Australian sovereign PGM program.

The Unique Australian Context

Maintaining control over the production and sustainment of essential defence capabilities is an existential concern for any state.[25] Sovereign defence acquisition has therefore been a preoccupation for statesmen and scholars dating back to the origins of the nation-state

---

[25] The classical statement is the Melian Dialogue in Thucydides's *History of the Peloponnesian War*; modern statements are Morgenthau and Waltz on realism. See Robert B. Strassler, *The Landmark Thucydides: A Comprehensive Guide to the Peloponnesian War*, New York: Free Press, 1998; Hans J. Morgenthau, *Politics Among Nations: The Struggle for Power and Peace*, New York: Knopf, 1948; and Kenneth N. Waltz, *Theory of International Politics*, New York: McGraw-Hill, 1979.

itself.[26] Contemporary scholarship on the subject, in the subfields of defence economics and defence innovation, emphasises the singular nature of national defence acquisition systems.[27] Every such system is unique, because each is subject to distinctive strategic, political, social, cultural and economic dynamics at the national, institutional and programmatic levels.[28] No two nations confront exactly the same strategic problems, nor do they possess precisely the same resources or assets. National institutions, processes and norms are the products of different histories and particular circumstances. As a result, despite apparent similarities in functional purposes and forms across states, each national defence acquisition system is bespoke to the nation it serves. Each looks and works a bit differently.

Care must therefore be taken in attempting to apply insights or 'lessons' across national defence acquisition systems. This may be particularly applicable to Australia's circumstances. Australia's strategic problems, situated at the fulcrum of the emerging Pacific Century, are unique in the international system.[29] Australia's defence-related politics, its systems and institutions for defence acquisition, and its technological strengths and weaknesses are all distinct from those of even its closest allies.[30] Sovereign defence acquisition will therefore need to be tailored to the Australian context.

It is also important to note that guided weapons are significantly different from many other defence capabilities and will therefore require unique approaches. For example, when comparing cost categories for surface ships and tactical missiles (as proxies for guided weapons), based on data from the U.S. Cost Assessment and Program Evaluation office in the Office of the Secretary of Defense, we see stark differences in the life-cycle cost categories. As shown in Figure 1.3,

---

[26] One of Machiavelli's central tenets in *The Prince* is that the state must control essential military capabilities rather than hiring them from abroad; see Azar Gat, *A History of Military Thought*, Oxford: Oxford University Press, 2001, p. 5. Tilly also famously observes that the financial and institutional demands of developing early modern weapons drove the development of the Western nation-state. Charles Tilly, *Coercion, Capital, and European States, AD 990–1992*, London: Blackwell, 1992.

[27] In the defence innovation subfield see, for example, Tai Ming Cheung, 'A Conceptual Framework for Defence Innovation', *Journal of Strategic Studies*, June 2021; in defence economics see, for example, Sabrina T. Howell et al., 'Opening Up Military Innovation: Causal Effects of "Bottom-Up" Reforms to U.S. Defense Research', IZA—Institute of Labor Economics Discussion Paper No. 14297, April 2021.

[28] Eugene Gholz and Harvey Sapolsky, 'The Defense Innovation Machine: Why the U.S. Will Remain on the Cutting Edge', *Journal of Strategic Studies*, June 2021; Marc DeVore, 'Armaments After Autonomy: Military Adaptation and the Drive for Domestic Defence Industries', *Journal of Strategic Studies*, May 2019, pp. 325–359; Magnus Petersson, 'Small States and Autonomous Systems—The Scandinavian Case', *Journal of Strategic Studies*, December 2020; Richard Bitzinger, 'Military-Technical Innovation in Small States: The Cases of Israel and Singapore', *Journal of Strategic Studies*, June 2021; Tai Ming Cheung, Thomas G. Mahnken and Andrew L. Ross, 'Frameworks for Analyzing Chinese Defense and Military Innovation', in Tai Ming Cheung, ed., *Forging China's Military Might: A New Framework for Assessing Innovation*, Baltimore: Johns Hopkins University Press, 2014.

[29] Defence, 2020a, pp. 11–13.

[30] See, for example, Stefan Markowski and Peter Hall, 'Defence Procurement and Industry Development: Some Lessons from Australia', in Ugurhan G. Berkok, ed., *Studies in Defence Procurement*, Kingston, Ont.: Queens, 2006.

FIGURE 1.3

Costs by Type for Tactical Missiles Versus Surface Ships

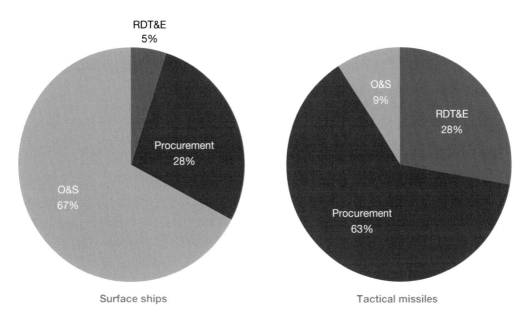

Surface ships   Tactical missiles

research, development, test and evaluation (RDT&E) represents over one-quarter of total life-cycle costs for tactical missiles, as opposed to just 5 per cent for surface ships. The figure also shows that procurement represents about two-thirds of life-cycle costs for tactical missiles, as opposed to about one-third for surface ships.[31] Just as an Australian Sovereign Guided Weapons Enterprise cannot directly mirror similar programs in other countries, an approach to such an enterprise cannot directly mirror other sovereign Australian defence enterprises.

Methodology

The research team examined a range of case studies to identify options and highlight challenges that may be faced by Australia in establishing a Sovereign Guided Weapons Enterprise. The five selected international case studies are detailed in this main report. Two domestic case studies were also considered: the Australian Naval Shipbuilding Enterprise (ANSE) and Australia's munitions manufacturing industry (MMI). These two domestic case studies are detailed in Appendices A and B (not available to the general public). This main report also distils the overall lessons, both international and domestic.

For the international case studies, we examined a diversity of experiences, which, in turn, should help inform the spectrum of potential approaches to the enterprise. We conducted the

---

[31] Office of the Secretary of Defense, *Cost Assessment and Program Evaluation*, September 2020.

case studies through literature reviews designed to capture the broad contexts in which the sovereign weapons capabilities were developed, how partnerships between the governments and private sectors were established, the requisite industrial base requirements in each case, and how exports were or were not developed. From that broad examination, we synthesised lessons for consideration as Australia develops its Sovereign Guided Weapons Enterprise. We selected the range of case studies to capture both some countries with long-established guided weapons industries and other countries with recently established sovereign industries. Countries examined include those with proven success selling guided weapons on the competitive international marketplace and others with less success. The countries examined also have a range of policies associated with developing, procuring, operating and sustaining their defence systems. The two domestic Australian enterprise chapters synthesise prior RAND research on the ANSE and the Australian MMI.

As will be discussed in further detail, the Sovereign Guided Weapons Enterprise will need to be bespoke to the Australian domestic and strategic context. Consequently, this analytical effort, and its associated methodology, is *not* designed to define what Australia's guided weapons enterprise should look like or how it should be implemented. It is rather intended to contribute international and domestic experiences to the ongoing discussion and to help in exploring the broad range of options that could be considered. More research will be needed to examine preferred options, decision trade-offs, and the best next actions associated with establishing the enterprise.

Organisation of This Report

Chapters Two–Six describe the relevance of the five international case studies to Australia. The case studies—of comparable enterprises in Japan, the United Arab Emirates (UAE), the United Kingdom, Canada and Norway—review the most relevant aspects of each country's experience and its lessons for Australia.

Appendices A and B (not available to the general public) describe the relevance of the two domestic case studies to Australia. Appendix A focuses on the ANSE, while Appendix B focuses on Australia's MMI.

Chapter Seven concludes with a synthesis of the overall lessons, both international and domestic, for an Australian Sovereign Guided Weapons Enterprise.

CHAPTER TWO

Japan Case Study: Guided Weapons Manufacturing

This case study focuses on the manufacturing of three guided weapons in Japan. While Japan's unique constitutional constraints have shaped its defence industry, there are useful insights for developing a cost-effective portfolio, leveraging international partnerships, and optimising the role of government. This chapter also provides introductory background, plus insights on Japan's defence industrial base, guided weapon export markets, and lessons for Australia.

Background

Japan finds itself in an increasingly challenging security environment in 2021, with two longtime threats—from both the Chinese Coast Guard's encroachment into Japanese territorial waters near the Senkaku Islands and the North Korean nuclear and ballistic missile programs that can reach Japan—growing more severe. These two countries are seen as the greatest threats to Japan's security.[1] Japan also sees Russia as another challenge in the Indo-Pacific. Due to this challenging environment, Japan looks to both strengthen its own self-defence capabilities and partner with other countries, with an emphasis on its alliance with the United States.

The *National Defense Program Guidelines* (NDPG), a core document in Japanese defence planning, describes the plan for Japan's defence force over the next ten years.[2] In that document for FY 2019 and beyond, Japan states that, as a peace-loving nation, it will continue to maintain an exclusively defence-oriented policy that will not threaten other countries. To meet this goal, the NDPG defines the following national defence objectives:[3]

- Create a security environment favourable to Japan by drawing on the strengths of the nation.

---

[1] Japan Ministry of Defense (Japan MOD), *Defense of Japan*, 2021.

[2] Japan MOD, 2021.

[3] Japan MOD, *National Defense Program Guidelines for FY 2019 and Beyond (Provisional Translation)*, 18 December 2018.

- Deter threats from reaching Japan by making adversaries recognise that doing harm to Japan would be 'difficult and consequential'.
- Counter threats that reach Japan to minimise potential damage.

To achieve these objectives, Japan plans to enhance capabilities for cross-domain operations. This effort aims to strengthen capabilities in new domains, such as cyber and space, while also adding to capabilities in traditional domains, such as maritime, air and missile defences. Ensuring that there are sufficient supporting capabilities—in the forms of a strong industrial base, logistics (fuel and infrastructure, for example) and other capabilities—is also highlighted as necessary for Japan's defence.[4]

Defence Budgets

Defence-related expenditures in Japan have risen, in nominal terms, for nine straight years, with FY 2021 expenditures planned to be 1.1-per cent higher than those of the previous year, at ¥5.12 trillion (roughly AUD 62 billion at publication) for FY 2021. However, the rate of growth in Japan's defence budget averaged only 2.3 per cent annually from FY 1995 to FY 2021, compared with average annual growth of 5.1 per cent for Australia and 4.2 per cent for South Korea during the same years.[5] As a percentage of gross domestic product (GDP), Japan's defence spending has remained below 1 per cent in recent times (reaching 0.94 per cent in FY 2020), although Prime Minister Yoshihide Suga has signalled a willingness to move beyond this threshold, given the current security environment.[6]

In breaking down the amounts spent in various categories of the FY 2021 budget, 43 per cent went to personnel and food provision, while the remaining 57 per cent went to material expenses. In a breakout of the expenditure by purpose, 17.9 per cent was directed to procurement of equipment and 2.2 per cent to R&D.[7]

Overall Defence Industrial Base

The Japan MOD considers the defence industrial base to consist of the 'human, physical, and technological bases that are essential for the production, operation, sustainment, and maintenance of defence equipment for the MOD/SDF's activities'.[8] Companies form much of the base, with defence being a relatively small fraction of the overall business for many of the major primes. Japan's largest two defence contractors, Mitsubishi Heavy Industries (MHI)

---

[4] Japan MOD, 2018.

[5] Calculations assume figures published by the government of each country are converted into USD using 2020 purchasing power parity; Japan MOD, 2021.

[6] Miki Nose, 'Politics: Japan Defense Spending Isn't Bound by 1% GDP Cap, Suga Says', *Nikkei Asia*, 13 August 2021.

[7] Japan MOD, 2021.

[8] Japan MOD, 2021.

and Kawasaki Heavy Industries (KHI), both have defence revenues totalling less than 10 per cent of their total revenues.[9]

Much of the defence industry remains concentrated in the hands of a small number of firms, especially MHI and KHI.[10] Six Japanese companies were listed in the Stockholm International Peace Research Institute (SIPRI) list of the top 100 companies for arms sales in 2018: MHI, KHI, Fujitsu, IHI Corp., Mitsubishi Electric Corp. and NEC Corp. Combined arms sales of these companies this year were estimated at USD 9.9 billion, with domestic defence sales being the sources of the overwhelming majority of that revenue.[11]

The Japan MOD has cited a few challenges associated with the current defence industrial base; the first is the high costs of domestic defence acquisition. These high costs are attributed to 'low volume, high-mix production' and a 'lack of international competitiveness'.[12] A second challenge relates to the somewhat pacifist or anti-militaristic nature of the Japanese population and its effect. Companies involved in defence are reluctant to highlight those aspects of their businesses for fear of being labelled as 'merchants of death' and seeing their commercial sales reduce.[13] As a result, these companies are less willing to advertise their defence systems broadly.

Another challenge for Japan's defence industry is the increasing fraction of the military acquisition budget spent on foreign military sales. Ninety-seven per cent of those purchases were supplied by the United States from 2016 to 2020.[14] The fraction of the total defence budget spent on U.S. foreign military sales climbed from 0.9 per cent in FY 2011 to 8.9 per cent in FY 2020, a fact more dramatic when considering that only 57 per cent of the budget goes to material expenses (and even less on capital expenses).[15] This trend toward buying more defence systems from the United States has cut into the amount of funding for domestic manufacturers, exacerbating some of the issues described previously.

Japan has undertaken several reviews targeted at strengthening its defence industrial base, with efforts intensifying in 2020 due to external threats and the impact of the COVID-19 pandemic. The focus of selected reviews has involved the examination of methods to increase domestic competition in the defence market and to strengthen the management of risks in supply chains. These methods will be discussed further in the following section.

---

[9] Alexandra Sakaki and Sebastian Maslow, 'Japan's New Arms Export Policies: Strategic Aspirations and Domestic Constraints', *Australian Journal of International Affairs*, Vol. 74, No. 6, 2020.

[10] Sakaki and Maslow, 2020.

[11] Aude Fleurant, Alexandra Kuimova, Diego Lopes Da Silva, Nan Tian, Pieter D. Wezeman and Siemon T. Wezeman, 'The SIPRI Top 100 Arms-Producing and Military Services Companies, 2018', SIPRI Fact Sheet, Solna, Sweden: Stockholm International Peace Research Institute, December 2019.

[12] Japan MOD, 2018.

[13] Daisuke Akimoto, 'Is Japan's Defense Industry in Decline?' *The Diplomat*, 1 October 2020.

[14] Fleurant et al., 2019.

[15] Rieko Miki, 'The Price of Peace: Why Japan Scrapped a $4.2bn US Missile System', *NikkeiAsia*, 5 August 2020.

Policies Related to Defence and Weapons Acquisitions

From at least the 1960s until 2014, Japan had a long-standing 'virtual ban' on exporting arms to other countries. This ban became institutionalised in 1967, when then Prime Minister Satō Eisaku asserted that Japan would not export arms to the Communist bloc, to countries subject to United Nations (UN) Security Council embargos, or to countries involved in or likely to be involved in international conflict.[16] Over time, these principles were gradually understood to apply more broadly, such that any kind of arms exports from Japan would be banned. Some exemptions to these restrictions were allowed over time, especially for working more effectively in developing and producing equipment with the United States. Such exemptions were seen as necessary to help respond to U.S. demands that Japan aid with 'burden sharing'.

In April 2014, then Prime Minister Shinzo Abe overturned the weapons export ban to help Japan play a larger role in countering China's growing military in the region. The move was seen by many as a way to allow Japan to provide military aid to Southeast Asian countries to help build their capacity to counter Chinese aggression in the South China Sea. American officials were supportive of lifting the ban, not just to help with regional security but also because the change in policy would ease Japan's participation in large, multinational military programs.[17] As a result of the ban, Japan's share of the global arms market between 1981 and 2011 had been only 0.13 per cent.[18]

The 2014 policy change that allowed export of defence equipment under a broader range of conditions is known as 'The Three Principles on Transfer of Defense Equipment and Technology'. The overall principles are quite detailed, but a summary of them highlights the types of transfers that are still prohibited, including those that violate international treaties or UN Security Council resolutions.[19] Exporting equipment or technology that is destined for a country that is party to a conflict is also usually prohibited. Transfers may be allowed under 'strict examination and control' when they contribute to promotion of peace, promotion of international cooperation, or the security of Japan. Permission must also be secured from the government of Japan before a nation transfers exported equipment or technology to a third party.[20]

In 2015, Japan stood up the Acquisition, Technology and Logistics Agency (ATLA) within the Japan MOD to foster a domestic defence industrial base and to enhance Japan's deter-

---

[16] Sakaki and Maslow, 2020.

[17] Martin Fackler, 'Japan Ends Decades-Long Ban on Export of Weapons', *New York Times*, 2 April 2014.

[18] Sakaki and Maslow, 2020.

[19] Japan National Security Council, 'Implementation Guidelines for the Three Principles on Transfer of Defense Equipment and Technology', 1 April 2014.

[20] Japan Ministry of Foreign Affairs, 'Three Principles on Transfer of Defense Equipment and Technology', webpage, 6 April 2016.

rence capability technologies. ATLA's website lists five missions that it pursues in support of this overall objective:

- ensuring technological superiority and responding to operational needs smoothly and quickly
- efficient acquisition of defense equipment (project management)
- strengthening of defense equipment and technology cooperation with other countries
- maintain and strengthen defense production and technological bases
- cost-reduction efforts and strengthening of inspection and audit functions.[21]

More recently, in its 2021 *Defense of Japan* document, the Japan MOD identified three goals for maintaining its defence industrial base: (1) ensure the sovereignty of security, (2) contribute to increased deterrent capability for the nation while improving bargaining power, and (3) improve the sophistication of domestic industry in working with cutting-edge technology.[22] To enact those goals, the government states that it will build long-term relationships between the public and private sectors while becoming more internationally competitive in the defence market.

When it comes to acquiring defence equipment, multiple mechanisms are utilised. On the one hand, the NDPG states that Japan's defence industry is essential for the production, operation and sustainment of defence capabilities. On the other hand, the NDPG highlights that procuring U.S.-made defence equipment will aid with interoperability with U.S. forces as part of alliance responsibilities.[23] Consequently, the Japan MOD pursues a variety of acquisition methods, including domestic development, international joint development and production, licensed domestic production, utilisation of commercial off-the-shelf products, and imports.[24] Each of these methods has different implications for the cost of delivering military capability and the current and future strength of Japan's domestic industrial base.

In maintaining and strengthening its defence industrial base, the Japan MOD has recently promoted the following measures:[25]

- improvements in contracting, as enacted by the 2015 Long-Term Contract Act
- initiatives in R&D
- initiatives for defence and technology cooperation
- building robust production and technological bases by better understanding supply chains
- strengthening Japan MOD functions in the industrial base via ATLA
- coordinating across government with other ministries and agencies.

---

[21] ATLA, 'Missions of ATLA', webpage, undated.

[22] Japan MOD, 2021.

[23] Japan MOD, 2018.

[24] Japan MOD, 2021.

[25] Japan MOD, 2021.

ATLA has the lead responsibility for understanding the defence supply chain and its associated risks. ATLA also manages the participation of Japanese companies in joint ventures with other countries, such as the F-35 program.[26]

The F-35 exemplifies the challenges faced when deciding the proper acquisition model for complex weapon systems, especially the price sensitivity of such decisions. Japan originally had planned to have MHI perform final assembly and checkout (FACO) locally for any F-35 aircraft ordered from 2013 onward.[27] But that decision was reversed in FY 2019, when analysis showed that the costs of local FACO were around USD 33 million higher per aircraft than the costs of importing completed F-35 aircraft.[28] To improve efficiency and save money, the government decided to purchase fully assembled F-35 aircraft starting in FY 2019.[29] However, that decision was reversed once again in 2020 as the costs of performing FACO had fallen, making local performance preferable once again. The latter decision was likely motivated by the increased size of the 2020 buy (105 new aircraft) and the fact that local FACO would be beneficial to Japan's efforts in developing a next-generation fighter aircraft.[30]

Another area of potential interest for increasing defence industry exports would involve the government provision of low-interest loans. Such a move would align Japan with other defence-exporting countries and would make purchases of equipment by developing countries more feasible. In the near term, the move is thought to be linked to improving Tokyo's bid to supply the Indonesian Navy with its 30FFM-class frigate.[31]

Guided Weapons

Japan builds several of its guided weapons domestically, with those listed in ATLA's defence industry pamphlet also listed in Table 2.1.

Three of the six systems in Table 2.1 will be covered in more detail below, but first we present a few overall observations on Japan's guided weapons programs. As discussed previously with defence acquisition in general, Japan uses a variety of methods to acquire guided weapons for its Self-Defense Force beyond domestic production. Some guided weapons are bought from foreign providers; these include the AIM-120C air-to-air missile manufactured by Raytheon in the United States and the Joint Strike Missile manufactured by Kongsberg in

---

[26] Japan MOD, 2021.

[27] Mike Yeo, 'Japan to cease in-Country Assembly of F-35 Jets', *DefenseNews*, 18 January 2019.

[28] Jon Grevatt, 'Japan Pays a Premium for Locally Built F-35s', *Jane's Defence Weekly*, 14 March 2018.

[29] Japan MOD, 2021.

[30] Jon Grevatt, 'Japan Commits to Local F-35 Production', *Jane's Defence Weekly*, 30 July 2020b.

[31] Jon Grevatt, 'Japan Looks to Introduce Finance System for Defence Experts', *Jane's Defence Industry*, 18 May 2021.

TABLE 2.1

Overview of Japan's Domestically Produced Guided Weapons

System Name	Manufacturer	Purpose
Middle-Range Multipurpose Missile	KHI	Road-mobile precision strike of surface targets
Type-11 Short-Range Surface-to-Air Missile	Toshiba Infrastructure Systems & Solutions Corporation	Countering short-range cruise and air-to-surface missiles
Type-03 Medium-Range Surface-to-Air Missile (Improved)	Mitsubishi Electric Corporation	Countering air threats, including cruise missiles
Type-12 Surface-to-Ship Missile	MHI	Land-based system to counter surface naval threats
Air-to-Ship Missile (ASM-3)	MHI	Supersonic cruise missile built to counter high-end warships
SM-3 Block IIA	MHI (joint development and manufacture with Raytheon in the United States)	Ballistic missile defence

SOURCE: Adapted from Japan MOD, undated.

Norway.[32] Some guided weapons are codeveloped with other nations; these include the SM-3 Block IIA (to be discussed next) and the Joint New Air-to-Air Missile (JNAAM), which is being co-developed with the United Kingdom. The JNAAM is a beyond-visual-range weapon and is Japan's first defence equipment program in recent history involving a partner besides the United States.[33] Japan is also investing locally in developing hypersonic weapons, specifically the Hyper-Velocity Gliding Projectile and a Hypersonic Cruising Missile. These weapons are intended to strengthen defence of Japan's 'remote islands', with deployment planned for the 2024–2028 timeframe.[34]

Ballistic Missile Defence: SM-3 Block IIA Joint Program

Japan must be prepared to defend itself from multiple threats in the air domain, including ballistic missiles, cruise missiles and aircraft that could come from North Korea or China. The threats from North Korea's nuclear and ballistic missile programs are of great concern. Therefore, air and missile defence capabilities are high priorities for the nation.

---

[32] Valerie Insinna, 'State Department Clears $113M Sale of AMRAAM Missiles to Japan', *DefenseNews*, 5 October 2017; and Gabriel Dominguez, 'Japan Awards Kongsberg Another Follow-On Contract for Joint Strike Missiles', *Jane's Defence Weekly*, 1 December 2020.

[33] Kosuke Takahashi, 'Japan Moves Ahead with JNAAM Co-Development', *Jane's Missiles & Rockets*, 22 December 2020b.

[34] James Bosbotinis, 'International Hypersonic Strike Weapons Projects Accelerate', *Aviation Week Intelligence Network*, 15 June 2020.

Japan's current ballistic missile defence program is multilayered, with upper-tier intercepts handled by Aegis-equipped destroyers and the lower tier carried out by Patriot PAC-3 systems.[35]

Modernisation plans had called for a deployment of two Aegis Ashore systems to augment Japan's missile defence capability and to be operational in 2023.[36] However, the Aegis Ashore program was cancelled in 2020 due to increasing program costs and concerns about where the defensive missile boosters might fall as parts of intercepts.[37] Later that same year, the cabinet approved plans to purchase two Aegis-equipped guided-missile destroyers to replace the Aegis Ashore sites.[38]

Related to these modernisation programs is the cooperative U.S.-Japan development of the SM-3 Block IIA ballistic missile defence interceptors that can be launched from Aegis-equipped ships or Aegis Ashore. Joint development of the SM-3 Block IIA started in 2006 with partners Raytheon Missile Systems and MHI. MHI is responsible for the second- and third-stage rocket motors, steering control and the missile nosecone; Raytheon has responsibility for system development and overall integration.[39] Acquisition of the missile started with Japan's FY 2017 budget and deployment of the system is planned for FY 2021.[40]

Whereas official documents clearly highlight the success of the SM-3 Block IIA cooperation, detailed discussions with government officials highlight both potential benefits and drawbacks associated with this kind of joint development and manufacturing.[41] There is broad agreement about the interoperability benefits of two allies developing, manufacturing and operating the same missile systems. This is especially true if the allies are likely to employ the system in a manner, such as combined air and missile defence, that requires interoperability. For Japan's defence industry, there have also been benefits in terms of profits from the program and exposure to the industry processes of larger organisations; this exposure should improve program management and systems engineering. ATLA has also gained knowledge that should help it improve program management.

On the less positive side, it appears that sharing intellectual property (IP) was a concern of both partners during the program. Specifically, concerns were raised about how information would be used going forward. Such concerns, as well as Raytheon's extensive experience with

---

[35] Japan MOD, 2021.

[36] Gabriel Dominguez, 'US Department of State Approves Possible Sale of 73 SM-3 Block IIA Missile to Japan', *Jane's Defence Weekly*, 28 August 2019.

[37] Jon Grevatt, 'Japanese MoD Reviews Procurement Priorities', *Jane's Defence Industry*, 7 July 2020a.

[38] Kosuke Takahashi, 'Update: Tokyo Approves Plan to Develop Type 12 Missile into Stand-Off Weapon', *Jane's Defence Weekly*, 18 December 2020a.

[39] Franz-Stefan Gady, 'US State Department Greenlights $3.3 Billion Missile Sale to Japan', *The Diplomat*, 12 September 2019.

[40] Japan MOD, 2021.

[41] Japan MOD, 2021; and discussions with ATLA staff, Microsoft Teams, September 2021 (name withheld on request).

similar missile interceptor programs, may have limited the contributions of MHI and ATLA in the missile design.[42] Although Japan likely gained knowledge about the efficient manufacture of the system and of specific subcomponents, it seems that relatively less knowledge was gained to inform the design of a similar system in the future.

Upgrades to Anti-Ship Missiles: ASM-3A and Type 12

The ASM-3A and Type 12 anti-ship missiles highlight the ability of Japan's defence industry to deliver weapons to address some of their most critical technical challenges while addressing specific geographic challenges for Japan.

In early 2021, the Japanese MOD decided to mass-produce an extended-range version of its developmental ASM-3 missile. The envisioned ASM-3A will be an air-launched, supersonic anti-ship missile that should be well suited to the surface maritime threats faced in the region. The weapon is planned to reach a top speed of Mach 3 with midcourse guidance of the missile reported to occur via GPS. A radar will reportedly provide terminal guidance.[43] The ASM-3A is too large to fit into the weapons bay of the F-35, so it is being integrated onto the Japanese Air Self-Defense Force's F-2 fighter and is also planned to be carried by the future F-X fighter. The ASM-3A program, which is a successor to the domestically built Type 93 air-to-ship missile, is the result of domestic development by MHI.[44]

The Type 12 surface-to-ship missile is also undergoing upgrades. The Type 12 entered service with the Japan Ground Self-Defense Force in 2014, with launch from a ground-based mobile launcher. In December of 2020, the Prime Minister approved plans to extend the range of the Type 12 from roughly 200 kilometres to 1,000 kilometres, which will allow it to better defend Japan's remote islands from potential Chinese military actions. The Type 12 could likely be adapted to strike other surface targets as well, suggesting that Japan will need to make upcoming decisions about whether to develop capabilities to destroy ballistic missiles in enemy territory prior to launch.[45]

Export Markets

Japan's success in marketing defence products to other nations has been limited since the restrictions were eased on arms exports in 2014. Japan has agreed to sell advanced air defence radars to the Philippines at a price of USD 100 million, which is clearly a success story. However, many other attempted sales have either stalled or failed, including the Soryu-class

---

[42] Discussions with ATLA staff, September 2021.

[43] David Axe, 'Japan Readies Its Ship-Smashing Super-Missile', *Forbes*, 16 July 2020.

[44] Kosuke Takahashi, 'Japan to Begin Mass Production of New ASM-3A Supersonic Anti-Ship Missile', *Jane's Defence Weekly*, 4 January 2021.

[45] Takahashi, 2020a.

submarine to Australia, the US-2 amphibious search and rescue aircraft to India, and the P-1 submarine hunting aircraft to the United Kingdom. The lack of success in these cases is attributed to several factors, including a lack of experience in selling defence equipment and a higher price associated with Japan's defence systems.[46]

In terms of selling guided weapons to external markets, Japan has had little success beyond the SM-3 Block IIA joint program. Japan's products are of high quality, but also come with a high price tag that makes international sales challenging. Another concern expressed by Japanese companies revolves around the protection of IP in any joint development and manufacturing program or in the foreign sales of weapons. Such concerns were also expressed as part of working with U.S. companies in joint ventures, as mentioned above.[47]

To be more successful with defence sales in the international marketplace, it has been suggested that a clear industrial strategy is needed to financially support Japanese defence manufacturers, provide the defence industry a method to offset the risks commonly involved with such sales, provide some method of financial support to countries wishing to import Japanese arms, and develop methods to ensure IP is protected.[48] Each of these areas would require investment by the Japanese government.

Lessons for Australia

Japan is an alliance partner of the United States, as is Australia, but already has an established defence and associated guided weapons industry. Although Japan could likely source all its guided weapons domestically, we have shown that Japan employs several methods of acquisition, including domestic development and production, international joint development and production, and foreign military sales.

Domestic sales appear to be preferred by Japan for most acquisitions, in part to maintain a strong sovereign industrial base. A resilient industrial base will help Japan 'effectively respond to a challenging security environment'.[49] Developing and manufacturing weapons domestically also allows the government to customise any weapon's capabilities to best address Japan's threat environment and geography. On the other hand, the drawback of buying guided weapons domestically is that low-volume production results in relatively high unit costs for capability. This high cost then limits the competitiveness associated with sales in the international marketplace.

International joint development and production appears well suited for systems for which interoperability is a necessity. Air and missile defence is a perfect example for the United

---

[46] Purnendra Jain, 'Japan's Weapon-Export Industry Takes Its First Steps', *East Asia Forum*, 30 October 2020.

[47] Arthur Herman, *The Awakening Giant: Risk and Opportunities for Japan's New Defense Export Policy*, Washington, D.C.: Hudson Institute, December 2016.

[48] Sakaki and Maslow, 2020.

[49] Japan MOD, 2021.

States and Japan in East Asia, where the number of sensors and missile interceptors are likely to be in short supply. By codeveloping and manufacturing the SM-3 Block IIA, there are clearly benefits, either if limited supplies of the missile must be shared in the future or if complex joint military operations requiring a close integration of forces must be conducted. Japan has also hoped to gain expertise both in managing complex weapons development and manufacturing programs, and in designing related missile technologies as part of the SM-3 Block IIA program. Knowledge has been gained in these areas, but it seems less than what might have initially been expected. There have also been concerns about a loss of Japan's IP during the cooperative program, although no evidence has been found to substantiate those claims.

Three lessons seem salient for Australia's future Sovereign Guided Weapons Enterprise. *First, if Australia's defence industry is asked to domestically develop and deliver a relatively small number of high-technology weapons for the ADF, they will likely come with relatively high unit and sustainment costs.* While this approach would better ensure a sovereign defence capability and the security of supply chains, it is likely that domestic delivery would come at a price premium. Relatively high costs would also limit the attractiveness of these weapons in the competitive international marketplace.

Second, joint development with allies and partner nations is attractive for a couple of reasons. First, having common systems aids interoperability and can reduce localised wartime supply shortfalls. Second, each partner in the joint development team has an ability to learn from the other partners—if information is shared freely. That said, concerns have been expressed about whether information shared in partnerships might be limited or used inappropriately. Partners in such programs might share less knowledge to protect IP.

Third, foreign sales may be needed to maintain an affordable Sovereign Guided Weapons Enterprise. If such sales are needed, the government can and should play a role in promoting and supporting them. Such support could come in the form of low-cost loans to purchasing nations, financing to support foreign sales, and/or methods to reduce the risks involved in the sales.

CHAPTER THREE

United Arab Emirates Case Study: Missiles and Weapons

This case study focuses on the UAE-based EDGE Group, particularly on its Missiles and Weapons cluster. While the UAE is certainly unique, particularly due to the role of sovereign wealth funds, we draw potentially applicable lessons on the use of offsets and government investment to spur interactions between domestic and international companies. This chapter also provides introductory background and insights on government and private-sector collaboration; partnership, licensing and other production details; and export markets, as well as lessons for Australia.

Background

EDGE Group resulted from a merger of more than 25 public and private companies that took place in November 2019. At that moment, EDGE Group immediately became one of the largest military companies in the world, with SIPRI placing it in the top 25 military companies globally by revenue in 2020.[1] Twelve of the top 25 global military companies by revenue are in the United States, according to SIPRI, thereby placing EDGE Group among the top 13 non-U.S. military companies.

The company's origins, along with the policies that spawned it, date back to the 1990s. We find that offsets—specifically, mandatory investments in the UAE by foreign contractors—played a key role in developing both the UAE defence sector and the business environment that enabled EDGE Group's founding. These offsets generated the original investments that created some of the companies that were later folded into EDGE Group. But more importantly in the context of this case study, these offsets facilitated the technology transfer agreements that assisted in the design of sovereign weapons, including PGMs.

Of course, Australia discontinued its formal offsets policy in the 1990s.[2] Uncertainty regarding the extent to which the policy imposed additional costs on contracts and whether

---

[1] SIPRI, 'Global Arms Industry: Sales by the Top 25 Companies up 8.5 Per Cent; Big Players Active in Global South', press release, 7 December 2020.

[2] Joint Standing Committee on Foreign Affairs, Defence and Trade, 2015, pp. 25, 40, 42–43.

it achieved its aim of facilitating technology transfers and building sovereign capabilities plagued the program. For example, Australia purchased 75 F/A-18 Hornets from McDonnell Douglas in 1987 and decided to build them domestically (rather than in the United States) to grow the Australian defence sector.[3] To fulfil its offset obligations (30 per cent of the project cost, which totalled billions of dollars), McDonnell Douglas marketed an Australian defence vehicle internationally and used Australian companies to produce a significant amount of the aircraft components. But no durable benefits appeared to result from this agreement, as the arrangements between McDonnell Douglas and the Australian firms did not extend beyond the project. This failure may have contributed to the dissolution of the Australian offsets policy shortly thereafter.

Nevertheless, Australia strategically implemented and adjusted its subsequent policies to realise the goals that the offsets policy had been designed to achieve. Namely, the AIC program requires that large contracts contain plans for maximising the involvement of domestic companies.[4] While not a formal offsets policy, the AIC seeks to facilitate similar outcomes, such as Australian firms accessing global supply chains. Therefore, to the extent that the UAE provides a comparable case, its success with EDGE Group's Missile and Weapons Cluster affirms Australia's strategy of adapting its own offsets policies to meet its contextual needs. The UAE's experience with offsets similarly produced some setbacks and uncertainties, but revisions to these policies, designed to address the UAE's specific needs, produced success in the long term.

Today, EDGE Group is comprised of five clusters: Platforms and Systems; Missiles and Weapons; Cyber Defence; Electronic Warfare and Intelligence; and Mission Support. Each cluster includes a group of streamlined entities. We focus on the Missiles and Weapons cluster due to its development of guided munitions, but we also provide information regarding the larger company's creation and evolution. Not only do some of the entities within the cluster date back to the 1990s, but the early policies that appear to have enabled EDGE Group's founding have been directly relevant to the establishment of a sovereign missile program.

History

Despite being founded in 2019, EDGE Group traces its origins back to the 1990s. Having witnessed with dismay the ease of Iraq's invasion of Kuwait in 1990, the UAE began modernising its military and designed a strategy to develop a domestic defence sector.[5] The UAE integrated these military efforts into a larger program that aimed to create sustainable and diverse domestic economic growth. Specifically, the UAE sought to develop a knowledge-based economy, create jobs for Emirati nationals, and build a national defence sector capable of high-value exports.

---

[3] Kim Beazley, 'Innovative Boost to Exports from Defence Offsets', press release, 23 January 1987.

[4] Greg Combet, 'Launch of Australian Industry Capability Program', press release, 26 February 2008.

[5] Florence Gaub and Zoe Stanley-Lockman, *Defence Industries in Arab State: Players and Strategies*, European Union Institute for Security Policy Studies, Chaillot Paper No. 141, March 2017.

A core component of this strategy involved the UAE Offsets Group (UOG), which was founded in 1992.[6] The UOG developed and implemented the offsets program, or the arrangement by which companies selling defence goods to the UAE were required to reinvest part of the proceeds of the transaction in the UAE. These sorts of offsets are used in Canada, Israel, Norway and a host of other countries. The UAE required foreign companies to return 60 per cent of their procurement contracts to the UAE's economy through local partnerships, which might involve local production or joint ventures. The original offsets program did not favour a specific industry but instead used a market-led approach that allowed foreign companies to develop partnerships in a variety of sectors. This approach resulted in investments across healthcare, agriculture, shipbuilding and other sectors.

As the size of defence contracts became larger, the complexity of the offset agreements increased. The UAE heeded the advice of a 1997 paper by the U.S. Commerce Department's International Trade Administration that suggested allowing companies to fulfil their offset obligations through deposits into investment funds.[7] This strategy would allow for maximum investment flexibility and more-targeted investments.[8] At the time, the UAE was one of very few countries to implement the strategy. To manage the investments and contracts from the resulting offsets, the UOG oversaw the creation of a sovereign wealth fund. This fund created a great deal of momentum and, over the next decade, numerous companies across various industries. Another fund, the Mubadala Development Company, was established in 2004, and the UOG transferred some of its assets to this newer fund to remain agile and streamlined. Several mergers and numerous spin-offs followed.

In 2012, the UAE government created an industrial park to geographically consolidate UAE defence companies and create cluster economies.[9] The Tawazun Industrial Park (TIP) includes not only office and manufacturing space, but regulatory offices, services such as information technology (IT) support; and personal services such as housing, shopping and laundry facilities. TIP hosts not just EDGE Group entities,[10] but also international firms such as Saab and Raytheon Emirates. EDGE Group works with both companies, suggesting the industrial park is effective at either creating such connections or facilitating them after they are developed.

By 2014, the UAE and business leaders assessed that the Emirati defence sector had matured enough to form the Emirates Defence Industries Company (EDIC). Multiple companies—

---

[6] Amin Badr-El-Din, 'The Offsets Program in the United Arab Emirates', *Middle East Policy*, Vol. 5, No. 1, 1997.

[7] Daniel Pearl, 'Offset Requirements of Defense Deals Often Have Little to Do with Purchaser', *Wall Street Journal*, 20 April 2000.

[8] Shana Marshall, 'The New Politics of Patronage: The Arms Trade and Clientalism in the Arab World', Brandeis Crown Center Working Paper, 2012.

[9] Michael E. Porter, 'Clusters and the New Economics of Competition', *Harvard Business Review*, November–December 1998.

[10] For example, EDGE Group is currently working with Raytheon to integrate a laser weapon onto one of its military vehicles. See EDGE Group, 'Al Jasoor Joins Forces with Raytheon Emirates, EARTH to Integrate High Energy Laser Systems onto Rabdan Vehicles', webpage, undated.

plus entire state investment funds that had founded, grown and purchased other defence companies—merged to create EDIC. The primary companies were Mubadala Development Company, Tawazun Holding and Emirates Advanced Investments Group, but there were over ten entities in total. EDIC, which represented the UAE's initial effort at creating a single integrated defence platform, was a precursor to EDGE Group. EDIC purchased more defence companies, including CARACAL and Burkan Munitions, and streamlined the redundancies among the merged entities.

EDGE Group then became the product of the follow-on, large-scale merger in 2019, only a few years after EDIC had formed. EDGE group incorporated not only EDIC, but over 20 other UAE-based defence companies as well. The result was a very large organisation with annual revenue of over USD 5 billion and the ability to manufacture guided missiles, aerial bombs and small-calibre ammunition, ranking among the top 25 largest military companies in the world in 2020 less than a year after its formation.

Joint Ventures and Technology Transfer with Offsets

The UAE relied heavily on the offset arrangements to create, grow and merge the companies that ultimately joined to create EDGE Group. The arrangements involved foreign partners in Canada (training), Germany (technology transfers), Russia (planned fighter jet coproduction), South Africa (technology transfer, joint production), South Korea (technology transfer), Sweden (technology transfer) and others.[11] EDGE Group companies entered into production agreements with defence companies based in those countries, received advanced training from them, and managed technology transfers of their IP to domestic companies. The offsets program and sovereign wealth funds were essential components of this process. Multiple companies that eventually became part of EDGE Group, including some entities within the Missiles and Weapons cluster, had been formed in the 1990s and early 2000s through strategic investments and had grown in part because of the offset arrangements.

The offsets program has evolved to address the UAE's changing defence-sector needs. For example, in 2019—the same year as EDGE Group's founding—the current iteration of the UOG (now called the Tawazun Economic Council and focused on defence growth) inaugurated a program known as Sustain & Enhance Emiratization in Defence & Security (SEEDS). SEEDS is an innovative vehicle for foreign companies to meet their offset obligations through internships for qualified Emirati students or through job placements for Emirati individuals at mid- and high-level positions in international companies. This program facilitates a 'soft' transfer of skills from contractors to Emiratis who will ostensibly apply them to domestic companies later in their careers. Tawazun piloted the program with German defence company Diehl; subsequent partners include Lockheed Martin, Thales and Saab.[12] The offset guidelines

---

[11] Gaub and Stanley-Lockman, 2017.

[12] Agnes Helou, 'Saab Inks Deal with Tawazun to Bolster Sensor Technology Research in UAE', *Defense News*, 22 February 2021.

implemented in 2019 allow for more-explicit technology transfers, investments and contractual engagements (e.g., manufacturing agreements) to fulfil offset obligations as well.

Missiles and Weapons Cluster

EDGE Group's Missiles and Weapons cluster currently consists of eight companies (with their areas of expertise shown in parentheses):

- HALCON (precision-guided missile systems)
- LAHAB Military Services (demilitarisation, training, testing, engineering, etc.)
- LAHAB Defence Systems (medium- and large-calibre munitions and weapon systems, including mortars, aircraft bombs and artillery)
- LAHAB Light Ammunition (small-calibre munitions)
- AL TARIQ (kits to convert unguided aerial weapons into precision munitions)
- APT (low-velocity projectiles and smoke grenades)
- CARACAL (small-arms design and manufacturing)
- REMAYA (military-grade shooting ranges).

These companies are streamlined versions of the entities that merged or were absorbed over the preceding decades. Their services and products are designed to provide a single integrated defence platform for clients and to meet a diversity of needs. We describe below how offsets through joint ventures were critical to the development of EDGE Group's PGMs specifically.

Government and Private-Sector Collaboration

EDGE Group is a state-owned enterprise. It emerged from the merger and purchase of a collection of both public and private companies, a process described above. The UAE used its offsets authority and multiple sovereign wealth funds to nurture several of the companies that eventually joined to create EDGE Group. The UAE also purchased private companies to diversify its holdings and ultimately create an integrated defence company. EDGE Group is operated by the government but functions as a commercial entity.

Partnership, Licensing and Other Production Details

EDGE Group's Missiles and Weapons cluster contains two companies involved in PGMs: HALCON and AL TARIQ. These companies are the descendants of multiple other companies that relied on support from the sovereign wealth funds, and both companies grew using technology transferred from an international partner.

In 2012, for example, a South African state-owned defence company named Denel created a joint venture with Tawazun Holding (one of the UAE's sovereign wealth funds) and agreed that the joint venture would have majority Emirati ownership.[13] The joint venture, called Tawazun Dynamics, developed and manufactured PGMs and initially focused on producing a variant of Denel's Umbani PGM. The technology transfer from Denel was critical for Tawazun Dynamics, for it jump-started the joint venture's ability to manufacture PGMs.

Similarly, EDGE Group is currently marketing Al Tariq, a system designed by the entity AL TARIQ, to convert unguided missiles into PGMs. The entity had also been originally enabled by the Tawazun Holding joint venture. Moreover, technology that HALCON uses for PGMs relies on technology that had been jointly developed with Denel through a different company that had preceded HALCON, which was founded in 2017. Joint ventures with international partners produced other EDGE Group Missiles and Weapons cluster companies as well, including CARACAL.

Export Markets

The UAE is actively increasing exports of defence products. Consistent with this, EDGE Group, as did its predecessors, exports to a variety of sovereign markets. For example, NIMR, an EDGE Group entity that designs and manufactures military vehicles, has a variety of clients in the Middle East and North Africa.[14] NIMR has sold its products to Saudi Arabia, Algeria, Libya, Egypt, Somalia and Oman, among others. The company may export to countries with arms embargoes as well, as approximately 50 NIMR vehicles made their way to Libya through Jordan.[15] Separately, EDGE Group companies have sold products or have agreements with many countries around the globe, including Russia, Malaysia, Ukraine, Serbia and Belarus.

EDGE Group's Missiles and Weapons cluster is also successfully exporting its products. Particularly relevant to this case study, HALCON has an agreement to supply German contractor Rheinmetall AG with the SkyNight counter-rocket, artillery and mortar (C-RAM) missile system.[16] Rheinmetall plans to integrate SkyNight with its Oerlikon Skynex Air Defence System.

SkyNight is the UAE's first sovereign air defence missile, and the UAE was its first customer. SkyNight's demonstration is expected in approximately three years.[17] HALCON is developing and advertising additional PGMs, but they are new or untested products that have had little time in the market.

---

[13] David Donald, 'Denel and Tawazun Join Forces', *IHS: Africa Aerospace & Defence*, 21 September 2012.

[14] Gaub and Stanley-Lockman, 2017.

[15] Gaub and Stanley-Lockman, 2017.

[16] Jeremy Binnie, 'IDEX 2021: Halcon to Provide C-RAM Missile for Rheinmetall for Air Defence', *Janes*, 23 February 2021.

[17] Chyrine Mezher, 'UAE's First Air Defense Missile to Be Used on German Oerlikon Skynex', *Breaking Defence*, 24 February 2021.

Lessons for Australia

Based on the UAE's experience of developing its defence capabilities, we derive lessons that can inform Australia's pursuit of sovereign PGMs. Of course, there are limitations on how well the insights from the UAE can be transferred to Australia, as the countries are functionally different. Australia's economy, along with its military budget, is considerably larger, and the strategic and geographic contexts are very dissimilar. Moreover, the UAE's priority with EDGE Group always has been to create a broadly integrated defence company—rather than develop a specific capability. Nonetheless, two lessons for Australia emerge from our case study.

First, creating and continually adapting an offsets program enabled the defence sector growth and joint ventures that transferred critical technology. Although Australia discontinued its offsets policy, it can continue to implement policies (such as the AIC program) that encourage sovereign capability development. For instance, the UAE implemented an offsets program in the early 1990s but later adjusted it numerous times to address evolving needs and the global marketplace. The UAE offsets program enabled the development and growth of domestic defence companies and ultimately a massive merger between them, which formed EDGE Group. The program also facilitated joint ventures with international partners, and these joint ventures provided critical technology transfers, including IP transfers, related to PGMs.

Second, an industrial defence park nurtured the cultivation of sovereign defence capabilities through interactions between domestic and international companies and the growth of local supply chains. Using government investments, the UAE developed a defence industrial park that provided physical space for a suite of defence companies. This space offered business benefits for Emirati firms and facilitated agreements with international companies.

CHAPTER FOUR

UK Case Study: UK Team Complex Weapons

This case study focuses on the UK Team Complex Weapons (TCW), which is a public-private arrangement. Like Australia, the United Kingdom has placed heavy emphasis on ensuring the viability of a sovereign precision weapons industry; however, the UK effort has prioritised overall cost reduction. This chapter provides introductory background on the development of TCW, plus discussions of the group's current status, its remaining challenges, and lessons for Australia.

Background and Development

TCW is a public-private arrangement between the MOD and the European missile designing and manufacturing company MBDA. The arrangement is meant to guarantee both the viability of the UK complex weapon industry and the freedom of action (FOA) of UK armed forces.[1] The arrangement is rooted in the United Kingdom's 2005 Defence Industrial Strategy, in which the MOD expressed concerns about sustaining the UK complex weapons supply chain in the context of 'unmitigated open international competition'.[2] A key concern was the growing importance of precision-guided weapons to maintain a competitive advantage in warfare, coupled with their ballooning acquisition and support costs.[3] The ability of complex weapons to reduce collateral damage and to integrate into network-based capabilities was also acknowledged as important for the United Kingdom's 'operational sovereignty'.[4]

The MOD unveiled its first iteration of the TCW agreement in July 2006. It brought together some of the most prominent players in the UK complex weapons industry— Qinetiq, Roxel and Thales—under the leadership of MBDA. The government presented TCW as a way to provide MOD with 'increased value for money, [to provide] industry with long-term assurance over spending plans and to encourage competition in the supply

---

[1] MBDA Missile Systems, 'Team Complex Weapons', webpage, 2021b.

[2] MOD, *Defence Industrial Strategy*, Defence White Paper, December 2005.

[3] MOD, 2005.

[4] MOD, 2005.

chain'.[5] No explanation was given as to how TCW could simultaneously give greater assurance to industry while stoking competition between its players, but the House of Commons Select Committee on Defence welcomed the agreement as a realistic way to 'help sustain key [complex weapons] skills and capability within the UK'.[6] It is not clear if the lawmakers realised that achieving cost savings, the primary goal of the policy, inevitably results in job loss unless offsetting production increases also occurs.[7]

The TCW agreement was one of five partnering agreements proposed by the 2005 Defence Industrial Strategy (the others relating to rotary-wing aircraft, fixed-wing aircraft, armoured fighting vehicles and submarines, respectively). Remarkably, TCW was the only one that came to fruition, suggesting that the arguments regarding supply chain sustainability and cost-cutting were particularly suited to guided rockets and missiles.[8]

In 2008, an Assessment Phase (AP) was launched for TCW, with two contracts made separately with MBDA and Thales to develop the following six complex weapon projects:

- Indirect Fire Precision Attack Loitering Munition (MBDA)
- Future Air-to-Surface Guided Weapon (heavyweight) for Royal Navy helicopters (MBDA)
- Future Air-to-Surface Guided Weapon (lightweight) requirement (Thales)
- Selected Precision Effects at Range (SPEAR) requirement for fast jets and helicopters (MBDA)
- Common Anti-Air Modular Missile (CAMM) family, including Sea Ceptor—for a Future Local Area Air Defence System (FLAADS) for the T23 Frigate and the Future Surface Combatant (MBDA)—and Land Ceptor
- Storm Shadow upgrade (MBDA).[9]

In 2010, following the satisfactory conclusion of MOD's AP,[10] MBDA announced that it was entering into a long-term bilateral arrangement, called a Portfolio Management Agreement (PMA), with the MOD.[11] The PMA was designed to improve 'the UK's CW [complex weapons] capability through the management of a portfolio of projects potentially worth up to £4Bn over the next 10 years'.[12] As part of the PMA, MBDA was assured that it would be able

---

[5] Craig Hoyle, 'Farnborough: MBDA to Head UK Arms Team', *Flight Global*, 24 July 2006.

[6] House of Commons Select Committee on Defence, *Sixth Report*, 19 December 2006.

[7] House of Commons Select Committee on Defence, 2006.

[8] House of Commons Defence Committee, *Defence Equipment 2008: Tenth Report of Session 2007–08*, 27 March 2008.

[9] MOD, 'MoD Launches a New Approach to Acquiring Complex Weapons', press release, 25 July 2008.

[10] MOD, 'The Competition Act 1998, Public Policy Exclusion Order 2007 No. 1896: Complex Weapons', webpage, 11 January 2021b.

[11] The PMA was initially interim and then extended in 2013. See NAO, *The Major Projects Report 2013*, Ministry of Defence Report, HC 817-I, Session 2013–2014, 13 February 2014.

[12] MBDA Missile Systems, 'MBDA & UK MOD, Long Term Partnering for Complex Weapons', press release, 29 March 2010.

to compete on the export market without legal challenges based on charges of unfair competition related to its bilateral agreement with the MOD.[13]

Also in 2010, MBDA was awarded an initial £330 million contract for the management and delivery of three programs: the Fire Precision Attack Loitering Munition, SPEAR, and Sea Ceptor weapons. All three were consolidated into a single portfolio for a total announced value of £4 billion, with £2.1 billion of gross savings expected over the next ten years.[14]

In 2011, the PMA was sealed in the Through Life Enabling Contract, which ensured MBDA's ownership over all designated programs, regardless of their phase in the acquisition process.[15] The resulting slate of contracted projects, collectively known as the Complex Weapons Pipeline (CWP), appeared for the first time in the 2012 National Audit Office (NAO) report, under MOD acquisition records. The report described the CWP as a way to deliver 'improved, adaptable and flexible Complex Weapons that can be shaped to meet current and future military capability needs' and underscored its importance for maintaining the United Kingdom's FOA and operational advantage (OA) 'through a sustained indigenous industrial construct'.[16] In a report presented by the MOD to Parliament the same year, FOA and OA were reasserted as strategic objectives to be attained through technological development.[17] (However, the MOD has no legal or contractual obligations to commit to future work, and the MOD retains the option to procure outside the pipeline.)

The total revised value of MBDA's portfolio was nearly doubled in the 2013 NAO *Major Projects Report* to £7.7 billion for the 2009–2023 period, but the £2.1 billion in expected savings was maintained for the initially agreed ten-year period.[18] The report also mentioned noncalculated but expected export benefits for the programs 'through the commercial exploitation levy and through savings achieved as a result of economies of scale'.[19] The PMA was henceforth extended in 2013, and the MOD monitored the savings until 2020, when 99.6 per cent of the target was considered banked (see Figure 4.1).[20]

---

[13] MBDA Missile Systems, 2010.

[14] MBDA Missile Systems, 2010. According to the NAO, the £2.1 billion in gross savings yields £1.2 billion in actual savings after subtracting the notional additional cost of single-source procurement from the benefits of the current procurement strategy. NAO, *Major Projects Report 2015 and the Equipment Plan 2015 to 2025*, Ministry of Defence Report, HC 488-I, Session 2015–2016, 22 October 2015.

[15] MBDA Missile Systems, 'Achieving Benefits', webpage, 2021a.

[16] NAO, *Ministry of Defence—The Major Projects Report 2012*, HC 684, Session 2012–2013, 10 January 2013.

[17] MOD, *National Security Through Technology: Technology Equipment, and Support for UK Defence and Security*, February 2012.

[18] NAO, 2013.

[19] NAO, 2014.

[20] NAO, *The Equipment Plan 2016 to 2026*, Ministry of Defence Report, HC 914, Session 2016–2017, 27 January 2017.

FIGURE 4.1

Recorded Savings of the Complex Weapons Program, 2014–2019

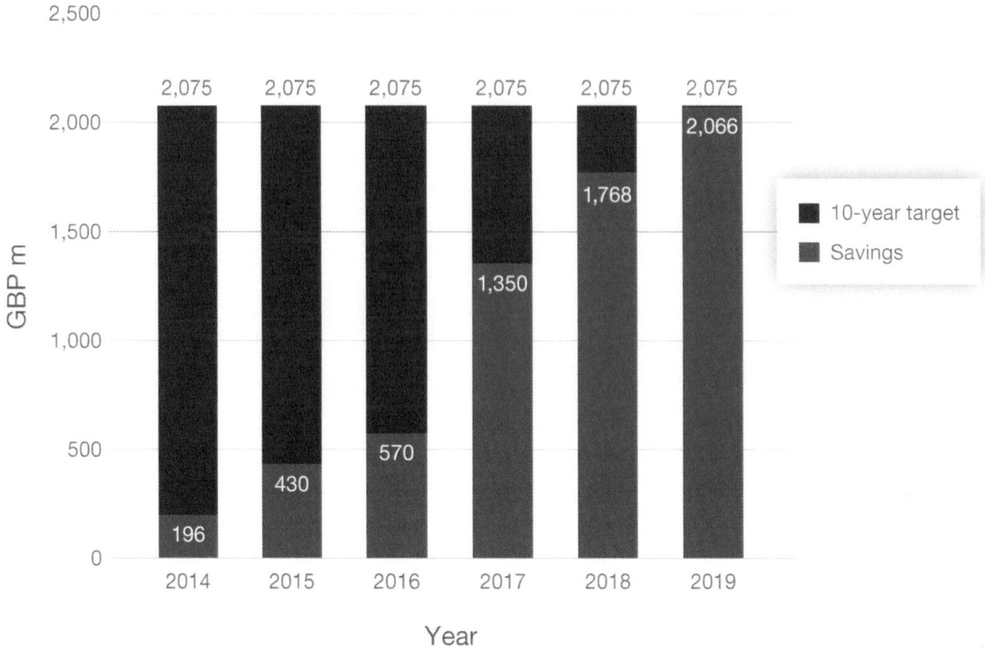

SOURCE: Adapted from MOD Equipment Plans, 2015 to 2020.

The financial benefits of the TCW were meant to be gained by using common and modular subsystems and technologies, working collaboratively, optimising designs to minimise through-life costs, and enabling greater flexibility to trade between requirements and costs across the portfolio. At the onset of TCW, just under half (47 per cent) of the financial benefits were expected to be reaped by sharing components among projects. TCW was also intended to benefit both the MOD and industry. The most significant benefit to industry was a clear pipeline of future work, which would help raise certainty and improve the industry's ability to plan for resources and sustain critical skills in the long term.[21]

Although not included in the financial benefits outlined above, the TCW/PMA approach was also expected to have other, broader benefits, including increased exports of UK-developed weapons and increased long-term sustainability of the UK domestic CW industry. Interestingly, the NAO also noted that at the beginning of the TCW, the £1.2 billion in financial benefits had already been 'banked' by the MOD and been included in its spending assumptions as part of its Equipment Plan. This structure meant that if projects in the pipeline were to be delayed or cancelled, some of these banked benefits could be lost (as

---

[21] NAO, 2014.

the MOD would either have to incur spending above budgeted levels or reduce spending elsewhere).[22]

Although one of the core benefits of TCW was meant to be better sustainment of critical complex weapons skills within UK industry, making such skills available to the MOD proved challenging in the early stages of TCW implementation. In 2012 and 2013, the NAO raised concerns about MOD's ability to provide capability and to oversee the portfolio of this new procurement approach, which required different skills and new ways of working. In May 2013, the MOD teams responsible for implementing TCW were 39 employees (18 per cent) short of their 217 full-time equivalent posts, which was recognised as a significant risk to delivery to both the MOD and its industry partners. The NAO also highlighted instances in which the MOD teams had relied too heavily on their industry partners because of MOD's resourcing challenges. As a result of these concerns, the MOD pursued a significant joint MOD/MBDA training program, including the development of a bespoke complex weapons portfolio skills training course in partnership with Cranfield University.[23]

Current State of Play

The MOD's current CWP encompasses a variety of complex weapon ranges, weights, launch domains and intended targets. Although some of these weapons are imported from the United States and other allied countries, the United Kingdom can produce many of the weapons domestically, mostly under MBDA's PMA, which currently covers the following weapon systems:

- anti-ship: Future Anti-Surface Guided Weapon (FASGW) (Heavy), also known as Sea Venom
- air-to-air: Advanced Short Range Air-to-Air Missile (ASRAAM), also known as Meteor
- air-to-ground: Brimstone, SPEAR3, Storm Shadow
- surface-to-air: CAMM, also known as Sea Ceptor and Land Ceptor.[24]

The development of systems within the CWP relies on partnerships with European allies, primarily France, to develop the capabilities for UK national defence. An example is the FASGW (Heavy), which is known as Sea Venom in the United Kingdom and as ANL (Anti-Navire Léger) in France.

Modifications and improvements under the PMA might involve the replacement of sub-system components; for instance, a Lightweight Multirole Missile (LMM) replaced an old

---

[22] NAO, 2014.

[23] NAO, 2014.

[24] MOD, *The Defence Equipment Plan 2019: Financial Summary*, 27 February 2020.

Roxel motor with a new Nammo motor.[25] Bringing in new suppliers, whether from the UK domestic market or internationally, can also help ensure that the UK complex weapon program retains peak performance. But even when an initial supplier is involved, capability sustainment programs can result in additions of new technologies, such as the Brimstone 2 dual-mode seeker, which improve weapon performance.[26]

The United Kingdom has also extended the lifespans, capabilities and subsystems of existing complex weapons through a program of 'spiral development', as evidenced by the contract awarded to MBDA to integrate SPEAR3 onto the F-35 Joint Strike Fighter (JSF). This integration of the SPEAR3 and the F-35, which builds on previous technology achievements, will start in 2021. The initial demonstration phase will consist of successive cycles of testing, simulation and trials.[27]

Upcoming Challenges

According to the United Kingdom's *Defence Equipment Plan 2019: Financial Summary*, MOD will spend £8.8 billion on complex weapon development between 2019 and 2029. Around 80 per cent of the budget will be for weapons delivered by MBDA through its PMA.[28]

The United Kingdom has a solid defence industrial base that makes the country a capable partner for its allies, but a significant technology gap exists between the United States and the United Kingdom. Russia and China are also ahead of the United Kingdom in investment levels, and both have world-class capabilities, particularly in technologies related to missile defence and hypersonics.[29]

Today, the UK complex weapons industry is trying to scale up with limited investments, and the current context of stagnating or declining budgets further limits industry growth and shrinks the competitive landscape.[30] Key future developments in areas such as missile defence, hypersonic systems, long-range cruise weapons or directed-energy weapons would require significant investments that would dwarf the cost of any individual weapon system brought into service in the past ten years.

Against this backdrop, the savings engendered by the PMA over the past decade warrant cautious optimism, but many of MOD's equipment programs have also gone over budget and

---

[25] Janes Weapon Systems database, *Air Launched Weapons.*

[26] Janes Weapon Systems database.

[27] MOD, '£550 F-35 Missile Contract Signed', webpage, January 6, 2021a.

[28] MOD, 2020.

[29] Paul Schwartz, *The Changing Nature and Implications of Russian Military Transfers to China*, CSIS, 21 June 2021.

[30] Schwartz, 2021.

beyond schedule.[31] Of the 32 programs currently defined as capabilities in MOD's Defence Major Projects Portfolio (DMPP), one-third face major risks to timely delivery, a situation exacerbated by funding shortfalls for equipment and support totalling £2.9 billion, which could grow to £13 billion should all the risks materialise.[32] The NAO has deemed the DMPP to be unaffordable, declaring that MOD 'has still not taken the necessary decisions to establish an affordable long-term investment program to develop future military capabilities' and 'has become locked into a cycle of managing its annual budgets to address urgent affordability pressures at the expense of longer-term strategic planning'.[33]

Lessons for Australia

The monetary savings associated with TCW were an essential design feature of the United Kingdom's sovereign capability. The savings were realised through economies of scale, specialised design, reuse of parts, use of common parts across builds, and other measures.[34] Ultimately, the up-front investments in sovereign guided munitions produced cost savings, since the program could be customised to the needs of the United Kingdom. Such an approach could similarly address the Australian concern about paying marginally higher costs for missiles because of supply chain adjustments.

Three lessons for Australia emerge from the UK case study. *First, developing custom guided munitions that are designed with cost-saving features, possibly with industry partnerships, can defray or eliminate the increased procurement costs of creating a sovereign program.* The United Kingdom used a combination of tactics, such as common parts across weapons and strategic management of stockpiles, to realise cost savings. These tactics appear to be especially suited to reducing costs in guided munition programs. Australia could consider commissioning multiple linked weapon systems that use modular parts and common design features.

The United Kingdom also leveraged the strengths of multiple defence companies to create a portfolio of guided munition systems. *Second, therefore, Australia could consider partnering with companies that have different strengths to create new guided munition systems.* This approach could help maintain the project's timing, help keep it on budget, and reduce the need to transfer IP to government-owned entities. This approach could also bring about a sovereign guided weapons program with broader capabilities through concurrent developments of air-to-ground, air-to-air and other solutions.

---

[31] NAO, *The Equipment Plan 2019 to 2029*, Ministry of Defence Report, HC 111, Session 2019–2020, 27 February 2020.

[32] NAO, 2020.

[33] NAO, 2020.

[34] Blake Stanton, 'Reducing Weapon Costs Through the Team Complex Weapons Portfolio Agreement', ATKINS, webpage, 6 February 2018.

Third, novel ways of partnering with industry or acquiring complex weapons may require new types of skills or additional resources within Defence. Without enough suitably qualified and experienced personnel to manage and provide oversight, Defence may risk relying too much on industry to deliver the intended benefits and fail to provide adequate challenges and assurances of what is achieved. Any potential implementation delays could result in delivery delays and subsequent cost increases due to inflation, disruption to industry, and intended benefits being lost.

CHAPTER FIVE

Canada Case Study: Munitions Supply Program

This case study focuses on Canada's Munitions Supply Program (MSP). This chapter provides introductory background on the program, along with discussions of its budget and policies, Canada's defence industrial base, its sovereign capability, and lessons for Australia.

Background

The Canadian Armed Forces (CAF) comprise some 66,000 active-duty personnel.[1] The Royal Canadian Navy operates 26 surface units and four submarines. Current acquisition programs will deliver up to 15 Canadian Surface Combatant (CSC) vessels, based on the Type 26 design, and six DeWolf-class Arctic Offshore Patrol Ships. The extant Halifax class frigates can operate Harpoon and Sea Sparrows, as well as the RIM-162 Evolved Sea Sparrow Missile (ESSM) anti-air missile. The new CSC ships will be fitted for SM-2MR Block IIIC, RIM-162 ESSM Block I and II surface-to-air missiles, and Tomahawk cruise missiles; MBDA Sea Ceptor air-defence missiles; and Kongsberg Naval Strike Missiles (NSMs). Canada's defence policy aspires for the Royal Canadian Navy to be able to deploy and sustain two task groups, each with up to four combatant ships, one supply ship and a submarine.[2]

The Royal Canadian Air Force has extended the life of combat-capable platforms of both the CF-18 Hornet fleet, which are complemented by former Royal Australian Air Force Hornets (to total 80 aircraft), and the CP-140 MPA. The primary weapons for the Hornets are the AIM-7, AIM-9, AIM-120C Advanced Medium-Range Air-to-Air Missiles (AMRAAMs) and a range of guided bombs. The Hornets should also be upgraded for the AIM-9X and the AGM-154 Joint Standoff Weapon (JSOW). A decision on the Royal Canadian Air Force's future fighter capability and new maritime patrol aircraft should be made by 2025. Canada will also acquire a remotely piloted Medium Altitude Long Endurance (MALE) aircraft, which should be able to deliver precision munitions.[3]

---

[1] Central Intelligence Agency, 'Canada', *The World Factbook*, last updated 14 September 2021a.

[2] Janes World Navies, 'Canada—Navy', webpage, last updated 13 May 2021b.

[3] Janes World Air Forces, 'Canada—Air Force', webpage, last updated 13 May 2021a.

Canadian Army guided weapons are limited to the Excalibur guided 155 mm artillery shells. The Canadian ground-based air defence capability was retired in 2012, and while the current defence policy calls for acquiring a new system, no program yet has been identified.[4]

Budget and Policies

Canada's defence spending has declined significantly since 1960, dropping below 1 per cent of GDP in 2014 at the time of Canada's withdrawal from Afghanistan, but then rising to 1.29 per cent by 2019.[5] This level remains well below the NATO guideline for member nations to maintain defence spending at 2 per cent of GDP.[6] While the 2017 Canadian defence policy[7] announced that defence investment would increase, Canada is unlikely to satisfy the NATO guideline or U.S. expectations.[8] This reluctance to increase defence spending to 2 per cent of GDP could damage any prospect of expanding today's U.S.-Canada defence industrial cooperation.[9]

A plan to increase defence spending in the 2017 defence policy (and for an increased pro-portion of the spending to be on capital investment) resulted from three trends: great power competition, the changing nature of conflict, and rapid evolution of technology.[10] An addi-tional motivation has been the need for modernisation, with ageing Canadian armed force weapon systems becoming comparatively less capable. Even with the increased investment, critics have noted that the majority of the planned spend is positioned in the second half of the 2017–2027 plan and that even this commitment is expected to raise military spending to only 1.4 per cent of GDP. In 2020, Canada's defence budget increased to 1.45 per cent; how-ever, this increase has been attributed to the decline in GDP caused by the COVID-19 pan-demic, rather than an increase in defence spending.[11]

The 2017 defence policy is based upon three elements: the strength of Canada's military to protect the country's sovereignty and to respond to domestic tasks; security through Canada's relationship with the United States, especially in the North American Aerospace

---

[4] Janes World Armies, 'Canada—Army', webpage, last updated 7 May 2021.

[5] World Bank, 'Military Expenditure (% of GDP)—Canada', webpage, undated a.

[6] NATO, 'Funding NATO', webpage, 13 August 2021b.

[7] Government of Canada, *Strong, Secure, Engaged: Canada's Defence Policy*, 2017.

[8] Amanda Connolly and Kerri Breen, 'Canada "Not on Course" to Hit 2% Defence Spending Pledge: U.S. Official', *Global News*, 16 February 2020; and Alan Freeman, 'Minister: Canada Will Build Up Its Mili-tary as the U.S. Pulls Back from World Stage', *Washington Post*, 6 June 2017.

[9] William Greenwalt, *Leveraging the National Technology Industrial Base to Address Great Power Competi-tion: The Imperative to Integrate Industrial Capabilities of Close Allies*, Atlantic Council Scowcroft Center for Strategy and Security, April 2019.

[10] Government of Canada, 2017, pp. 50–56.

[11] Lee Berthiaume, 'Canada Jumps Closer to Military-Spending Target Thanks to COVID-19's Economic Damage', *CCTV News*, 21 October 2020.

Defense Command (NORAD); and global engagement, with the military being able to contribute to NATO, coalition operations and capacity building. Canadian defence policy is less focused on major combat operations, with engagement tending more toward peacekeeping, and with the NORAD relationship not extending to active Canadian involvement in missile defence. With climate change providing greater access by other nations to the Northwest Passage, the sovereignty of Arctic territory is a key priority for Canada.[12] Arctic security is of such importance to Canada that it is the first of six Key Industrial Capability (KIC) themes, which are designed to strengthen the domestic defence industrial base.[13]

Defence Industrial Base

Canada's defence industry generated over CAD 10 billion (AUD 11 billion) in sales in 2016, contributing CAD 6.2 billion (AUD 6.8 billion) to GDP and providing over 60,000 jobs.[14] Much of Canada's defence industry works within global supply chains, especially in support of U.S. prime contractors. Integration with U.S. industry is facilitated through long-standing bilateral defence agreements, with Canada considered to be part of the U.S. Defense Industrial Base for production planning purposes.[15] Since the late 1950s, the Defense Production Sharing Arrangement (DPSA) has benefited both nations by fostering freer trade in defence materiel, providing Canada favourable access to the much larger U.S. military market.[16] Established shortly after the NORAD pact, the DPSA provides Canadian firms with access to U.S. contracts, thus averting the collapse of a previously struggling industry.[17]

Canada was also the original partner of the U.S. National Technology and Industrial Base (NTIB), which consolidated the North American defence industry during the drawdown of military spending at the end of the Cold War. The NTIB helps increase Canada's defence-related export revenue and provides relatively seamless supply chain integration, which allows one partner to surge to meet the needs of the other. This seamlessness is facilitated by Canada's replication of U.S. export control mechanisms, allowing supply interactions under exemption from the International Traffic in Arms Regulations (ITAR).[18] One perceived negative is that the associated commitment to U.S. ITAR obligations could limit other export

---

[12] Government of Canada, *Statement on Canada's Arctic Foreign Policy*, 2010, p. 2.

[13] Government of Canada, *Canada First: Leveraging Defence Procurement Through Key Industrial Capabilities*, February 2013, pp xiii–xiv.

[14] Innovation, Science and Economic Development Canada, *State of Canada's Defence Industry*, 2018, p. 4.

[15] Acquistion.gov, 'Part 225—Foreign Acquisition', webpage, undated.

[16] William M. Hix, Bruce Held and Ellen M. Pint, *Lessons from the North: Canada's Privatization of Military Ammunition Production*, Santa Monica, Calif: RAND Corporation, MG-169-OSD, 2004, p. 10.

[17] Donald Barry and Duane Bratt, 'Defense Against Help: Explaining Canada-U.S. Security Relations', *American Review of Canadian Studies*, Vol. 38, No. 1, 2008.

[18] Brendan Thomas-Noone, *Ebbing Opportunity: Australian and the U.S. National Technology and Industrial Base*, U.S. Studies Centre, November 2019, p. 7.

opportunities.[19] Additionally, the ITAR exemptions do not apply to a number of technology areas, including classified articles and missile technology.[20] Nevertheless, the close integration with the United States and the benefits enjoyed by Canada have not been able to be replicated by Australia or the United Kingdom, who joined the NTIB in 2016 but without ITAR exemption.[21]

The Canadian government supports the domestic defence industry primarily through its Industrial and Regional Benefits (IRB) framework, an offset approach that requires winners of major defence contracts to spend the equivalent of the dollar value of contracts in support of Canadian industry. The IRB framework also designates two special sectors—shipbuilding and munitions—for both demand-side and supply-side support.[22] For instance, the IRB has specific requirements for Canadian industry to supply these special sectors in a timely and secure fashion if that supply is not adequately provided by foreign contractors.

The munitions sector of firearms, small weapons, ammunition and rockets/missiles comprises around 5 per cent of Canada's defence products and services, with much of the related industry located in Quebec.[23] The bulk of this activity falls under Canada's MSP.

Munitions Supply Program

Canadian munitions production was critical to the allies in World War II. In subsequent decades, the munitions industry faced financial difficulties, with changing types and quantities of ammunition, a lack of export opportunities, and no funding to modernise facilities.[24] While a high proportion of the munitions industry was government-owned and government-operated up until the 1960s, from that time through the 1980s the so-called Crown corporations became privatised and commercially operated.

The MSP was established in 1978 to foster a domestic industry to provide self-sufficiency in the supply of high-volume-usage ammunition to Canadian forces. Preferred suppliers were identified, with the Department of National Defence (DND) providing certainty by issuing long-term plans and then negotiating annually costed agreements with suppliers.

There are currently five MSP participants: General Dynamics Ordnance and Tactical Systems Canada (GD-OTS-C), IMT Defence, Colt Canada, Magellan Aerospace and HFI Pyrotechnics. Each company provides a Centre of Excellence to the government of Canada for various aspects of munitions and small arms. Participants have access to export markets, with

---

[19] Government of Canada, 2013, p. 17.

[20] Andrew Hunter, Kristina Obecny and Gregory Sanders, *U.S.-Canadian Defense Industrial Cooperation*, Center for Strategic and International Studies, June 2017, p. 51.

[21] Thomas-Noone, 2019, p. 7.

[22] Government of Canada, 2013, p. 19.

[23] Innovation, Science and Economic Development Canada, *State of Canada's Defence Industry*, 2018, p. 16.

[24] Hix, Held and Pint, 2004, p. 33.

the proviso that priority is given to the Canadian government in times of urgent operational requirements.[25] However, surge requirements may also be satisfied by the DND stockpiling munitions or by each participant having an informal capacity to operate at above-normal levels.[26] Such informal arrangements are acceptable in the context of Canada's relatively benign security environment.

In 2007, a review of the MSP was undertaken by the Canadian Chief of Review Services (an audit function within the DND). The review questioned the relevance and impact of the program in the post–Cold War environment, especially with the lack of an industrial policy, the program supplying only land warfare munitions, significant offshore supply dependencies for many of the CAF's in-service systems, and the prospect for more cost-effective alternatives for trusted supply.[27]

Recognising continuing concerns that the program was outdated,[28] the DND in 2015 committed to reviewing and examining the policy, scope and governance of the MSP with an objective of reinvigorating it.[29] Following an extensive review, the MSP was reaffirmed in 2018, although the review documents are not public. The review did not revise any industrial policies other than putting a greater focus on innovation and on the greening of production systems.[30] Despite earlier concerns about the scope of MSP, the 2018 review did not consider any change to incorporate guided weapons, as there is no associated guidance in either the 2017 defence policy or the procurement policy (Canada First; see below).

Canadian industry does not have the capability for domestic development or production of guided weapons, with none of the MSP companies having expertise in the tightly controlled market of guidance systems. Canadian companies could produce components as part of a U.S. supply chain; however, any such initiative would primarily be economically driven, and as such it may not substantially improve the security of Canadian supply.

The main source for CAF guided weapons remains the U.S. government's Foreign Military Sales (FMS) program, with the Royal Canadian Navy also being part of the NATO Sea Sparrow Consortium for the development, procurement and maintenance of ESSMs. Canada continues to rely on foreign sources for guided weapons and missiles, primarily as the economics of the small quantities it requires does not justify a domestic solution.[31]

---

[25] Government of Canada, 2013, p. 20.

[26] Hix, Held and Pint, 2004, p. 56.

[27] DND, *Evaluation of Munitions Supply Program (MSP)*, Chief Review Services Document 1258-101-4, December 2007.

[28] Ugurhan Berkok and Christopher Penney *The Political Economy of the Munitions Supply Program*, Defence R&D Canada, Contract Report DRDC-RDDC-2014-C92, 2014.

[29] Government of Canada, 'Munitions Supply Program', webpage, last updated 28 April 2021.

[30] In this context, *greening* refers to the use of technologies and processes that reduce the impact of production systems on the environment.

[31] Hix, Held and Pint, 2004, p. 12.

Sovereign Capability

The Canada First defence strategy identifies the aforementioned KICs as a means of focusing government and industry efforts on a limited number of priorities. In a similar approach to Australia (having derived this policy from Australia and the United Kingdom), Canada got its KICs from considerations of security of supply, Canada-specific requirements, technology advances and interoperability.[32] Whereas high-volume small arms munitions are a KIC priority from a security-of-supply perspective, more-advanced munitions are not considered a priority. Consequently, the munitions KIC is focused on in-service support of high-volume small-arms munitions, and there are no KIC-related munitions entries in Canada's Defence Capability Blueprint.[33]

In addition to the previously noted economic argument against a sovereign weapons capability in Canada, there are two other factors that differentiate Canada's from Australia's situation. First, while appreciating the deteriorating security environment, the Canadian government sees its security risks as relatively low, especially with protections afforded by U.S. proximity, history and the NORAD relationship. Second, the supply routes from Canada's weapon suppliers, particularly the United States, are far less likely to be disrupted. Accordingly, the risks to supply of guided weapons do not justify a departure for Canada from trusted supply arrangements, such as FMS.

Lessons for Australia

While in some ways Canada is a like-minded nation to Australia, Canada's unique circumstances mean that it offers few direct lessons for Australia's development of a sovereign weapon industry capability. However, there are two indirect lessons that may be useful regarding integration with the U.S. defence industrial base and sustainment of surge capacity.

Canada's unmatched integration with U.S. industry is relevant to Australia's ambition of developing an advanced weapon capacity within its own domestic industry. While Australian industry might possess the innovation and enthusiasm to develop indigenous capabilities, a more viable approach may be to aspire to an environment in which domestic industry could coproduce weapons. An Australia-based production of guided weapons would still require substantial access to U.S. technology. Even with favourable ITAR exemptions, Canada has faced significant limitations in developing sensitive weapons; thus, it is reasonable to expect that Australia, absent similar exemptions, would face even greater limitations. As a result, *the first indirect lesson is that Australia's integration with U.S. industry and access to it would need to not only match the levels of integration and access already enjoyed by Canada, but surpass those levels.* This goal might seem unattainable. However, Australia's investment in defence

---

[32] Government of Canada, 2013, p. 25.

[33] Government of Canada, 'Defence Capability Blueprint', webpage, last updated 9 January 2020.

capability and its commitment to Indo-Pacific security may provide reasonable justifications. Furthermore, the recent AUKUS[34] and AUSMIN announcements of U.S. support to Australia's future sovereign weapons enterprise warrant greater confidence of the required access to advanced weapon technologies.

The second indirect lesson comes from Canada's approach to quantity planning within the MSP. A key problem in munitions planning is the friction between the uncertainty of operational conflict versus industry's need for certainty in capacity planning. Canada's munitions industry comprises single-sector companies that have limited abilities to pivot capacities from other activities toward munitions production when conflict arises. Canada's MSP has managed this challenge through five mechanisms: (1) a regime of notifying industry of requirements ahead of time, allowing for updates of plans on an annual basis; (2) a practice of stockpiling inventory, cognisant of item life; (3) an informal commitment by industry to maintain latent capacity that can be ramped up; (4) encouragement of export activities, which may be sacrificed if urgent Canadian requirements arise; and (5), although not formally part of the MSP, an understanding that under the NTIB, U.S. capacities could be directed toward urgent Canadian needs. *The second indirect lesson, therefore, is that while Canada has developed mechanisms to confer confidence in the case of its relatively less-complex munition requirements, such mechanisms may not suffice to produce more-complex weapon systems that involve more processes, supply chains and time.* Nevertheless, in the mature sovereign weapon enterprise, even complex weapons should be considered with a view to ongoing replenishment and inventory management, rather than a traditional acquisition mindset.

The Canada case study thus provides lessons in terms of how Australia might manage technology regimes in developing a sovereign weapon industry, as well as how Australia might address the key challenge of managing production to meet a potentially variable demand.

---

[34] AUKUS, a trilateral technology cooperation agreement between Australia, the United Kingdom and the United States, was announced on 15 September 2021.

CHAPTER SIX

Norway Case Study: Kongsberg Missile Programs

This case study focuses on Norway's Kongsberg missile programs. This chapter provides introductory background on the programs, along with discussions of their budget and policies, Norway's defence industrial base, and lessons for Australia.

Background

Norwegian Armed Forces consist of 23,000 active-duty personnel, not including specialist functions and the Home Guard.[1]

The Royal Norwegian Navy's major fleet units comprise four Fridtjof Nansen–class frigates and six Skjold-class corvettes. All of these vessels are equipped with the Norwegian-developed NSM. The frigates are also equipped with ESSMs, whereas the corvettes carry MBDA Mistral 2 missiles for short-range air defence.[2]

The Royal Norwegian Air Force is in the process of updating its combat and maritime surveillance capabilities from the F16 and P3C to the F35A (52 aircraft) and P8A (5 aircraft), respectively. Both new platforms will be capable of launching the Norwegian-developed Joint Strike Missile (JSM), with the F35A also to be equipped with AIM-9X and AIM-120C-7 missiles. The Royal Norwegian Air Force (and in due course, the Army) operate the National Advanced Surface to Air Missile System (NASAMS II) ground-based air defence system, a Norwegian-developed launching system that uses the AIM-120 missile against low-altitude air threats.[3]

---

[1] Central Intelligence Agency, 'Norway: Military and Security Forces', *The World Factbook*, last updated 10 November 2021b.

[2] Janes World Navies, 'Norway—Navy', webpage, last updated 14 April 2021a.

[3] Janes World Air Forces, 'Norway—Air Force', webpage, last updated 17 August 2021b.

Budget and Policies

Norway's defence spending was 1.68 per cent of GDP in 2019,[4] below NATO guidance but still higher than that of most NATO nations. Defence spending is estimated to rise to 1.85 per cent in 2021,[5] although, like in many nations, this is due more to a pandemic-induced reduction in GDP than to a rise in defence investment. Norway's defence budget is strongly committed to capability investment, with well above NATO's 20-per cent target allocated to such investment.[6]

Norwegian defence policy is built on three pillars: national defence capabilities, collective security through NATO, and domestic security through a Total Defence concept. The 2020 policy identifies the growing threats from several sources: great power rivalry, challenges to the rules-based world order, technological change, the broader use of instruments of power by nation-states, and a more demanding and more complex operational environment. These threats will lead to investments toward increased combat power for the Norwegian Armed Forces, with a plan to return to 2 per cent of GDP by 2028.[7]

Norwegian armed force capabilities are designed primarily to deter and defeat aggression against its interests. This commitment includes the prospect of Norway being drawn into a high-end conflict in support of NATO, potentially in conflict with Russia in its high north.[8] This risk demands that Norway invest in highly capable weapon systems.

Defence Industrial Base

Norway's exports of military arms in 2020 were worth AUD 1 billion, with the main customers being other NATO countries. This sum represents a 36-per cent increase over 2019, primarily due to increased purchases of air defence systems.[9] A key contributor to Norway's defence industrial base is the Kongsberg company. Across its maritime, aerospace and defence divisions, the company generated AUD 4 billion in revenue in 2020—also a significant increase over previous years.[10]

---

[4] World Bank, 'Military Expenditure (% of GDP)—Norway', webpage, undated b.

[5] NATO, 'Defence Expenditure of NATO Countries (2014–2021)', press release, 11 June 2021a.

[6] International Trade Administration, 'Norway—Country Commercial Guide', webpage, last updated 10 October 2021.

[7] Norwegian Ministry of Defence, *The Defence of Norway: Capability and Readiness, Long Term Defence Plan 2020*, 2020.

[8] James Black, Stephen J. Flanagan, Gene Germanovich, Ruth Harris, David Ochmanek, Marina Favaro, Katerina Galai and Emily Ryen Gloinson, *Enhancing Deterrence and Defence on NATO's Northern Flank, Allied Perspectives on Strategic Options for Norway*, Santa Monica, Calif.: RAND Corporation, RR-4381-NMOD, 2020.

[9] Government of Norway, 'Norwegian Exports of Defence-Related Products in 2020', webpage, 6 June 2021.

[10] Kongsberg, 'Key Figures 2020', webpage, undated b.

The Norwegian government fosters the national defence industrial base through a combination of R&D support, trusted and cooperative relationships between government and industry, long-term financial commitment, strong synergies between national requirements and export markets, and ongoing communication.[11] Through these mechanisms, the Norwegian defence industrial strategy supports Norway's industry in eight priority areas of technology competence, of which missile technology is one. The others are command and control (C2) systems, integration, autonomous systems and artificial intelligence (AI), underwater technology, ammunition and propulsion, materials technology and life-cycle support.[12] Although a successful defence industry contributes to the economy through jobs creation, the Norwegian defence industrial strategy does not emphasise the economic aspect, instead focusing upon national security needs.[13]

Given Norway's small local demand and the need for investment into core competencies, an export market is critical to the defence industrial base. The Norwegian defence industry's first and highest priority is servicing the needs of the country's armed forces, but there is also a view within the industry that Norway's use of its domestically produced weapon systems contributes to export demand.[14] The success of Norwegian defence companies in exporting their systems is aided by the companies' internal capabilities; their external collaboration, enabled especially by Norwegian public support for science and technology (S&T) research; and the external environment, including the Norwegian government's demand and offset policies.[15]

Norway has used offsets as a defence industrial policy since the middle of the last century. Arguably, this policy has been successful in achieving a range of security and economic motivations.[16] The Norwegian government also ensures the sustainability of major defence companies involved in weapons and munitions, namely Kongsberg and Nammo, through its ownership of controlling stakes in these commercial entities. This approach is designed to provide certainty that a sovereign industry will continue to provide for the national military without risking dependence on foreign suppliers.[17]

---

[11] Norwegian Ministry of Defence, *Cooperation for Security: National Defence Industrial Strategy, A Technologically Advanced Defence for the Future*, Meld. St. 17 (2020–2021), 12 March 2021, pp. 7–8.

[12] Norwegian Ministry of Defence, 2021, p. 12.

[13] Kjetil Hatlebakk Hove, *Defence Industrial Policy in Norway: Drivers and Influence*, Armament Industry Europe Research Group Paper 25, February 2018, p. 7.

[14] Fulvio Castellacci, Arne Fevolden and Martin Lundmark, 'How Are Defence Companies Responding to EU Defence and Security Market Liberalization? A Comparative Study of Norway and Sweden', *Journal of European Public Policy*, Vol. 21, No. 8, 2014, p. 1227.

[15] Fulvio Castellacci and Arne Fevolden, 'Capable Companies or Changing Markets? Explaining the Export Performance of Firms in the Defence Industry', *Defence and Peace Economics*, Vol. 25, No. 6, 2014, p. 552.

[16] Arne Fevolden and Kari Tvetbråten, 'Defence Industrial Policy—A Sound Security Strategy or an Economic Fallacy?' *Defence Studies*, Vol. 16, No. 2, 2016.

[17] Norwegian Ministry of Trade, Industry and Fisheries, *The State's Direct Ownership of Companies: Sustainable Value Creation*, Meld. St. 8 (2019–2020), 22 November 2019.

Kongsberg Missile Programs

Kongsberg is a company that has existed for over 200 years, with its weapons heritage dating to rifle production from the late nineteenth century.[18] Collaboration with public S&T organisations after World War II led to the production of innovative sovereign weapons that were subsequently exported, starting with the Penguin anti-ship missile. The Penguin missile also was a success for Australia, where Penguin warheads were built by Thales Australia under licence from Kongsberg. This Australian production was established as an offset for the Super Seasprite acquisition; however, following the demise of that program, the Australian-produced Penguin components were exported in the 1990s and 2000s.[19]

Work to develop a longer-range, low-observable successor to the Penguin began in the early 1990s. Ballistic tests of the NSM were undertaken in 2000, with guided tests beginning two years later. Various challenges during development and testing led Kongsberg to invest in the NSM program substantially beyond the government-contracted investment. Nevertheless, the Norwegian government committed in 2005 to fitting major fleet units with the NSM, a production contract was signed in 2007, and live ship firing of the production missile was first performed in 2012. Although Malaysia is the only other operator of the NSM, both the U.S. Navy and Canadian Navy have committed to its future use.[20]

With Norway as a partner in the JSF program, Kongsberg and Lockheed Martin indicated in 2007 their intention to jointly produce an air-launched version of the NSM (the JSM). The JSM is intended to have improved performance and range over the NSM, may be launched from platforms other than the JSF, and will complement the NSM's passive infrared seeker with a passive electronic support measures (ESM) system. Norwegian initial operational capability is expected in 2023, with other customers to include Australia, the United States, Canada, Japan and Italy.[21]

The development of the NSM distinguishes Kongsberg as a company that readily engages not only with Norwegian companies, such as Nammo, but with a range of Western defence companies. These companies have been engaged with Kongsberg in its missile supply chain (such as BAE Systems Australia for JSM seeker development) and its platform integration (such as Lockheed Martin). Raytheon has been a partner in both roles in support of the NSM and JSM. Kongsberg and Raytheon are also partners in the development and production of the AIM-120-based NASAMS air defence system.[22]

---

[18] Kongsberg, '200 Years of Determination', webpage, undated a.

[19] Corry Roberts, 'Australia's Sovereign Guided Weapons Heritage', *Thales*, 19 July 2021.

[20] Janes Weapons, 'Naval Strike Missile (NSM)', webpage, 17 February 2020a.

[21] Janes Weapons, 'Joint Strike Missile (JSM)', webpage, 6 July 2021.

[22] Janes Weapons, 'NASAMS', webpage, 19 November 2020b.

Lessons for Australia

Norway's defence industry strategy resembles that of Australia's, with the identification of key technology areas in which the government seeks to maintain technical competence. These technology areas were selected to provide benefits in terms of tailoring systems to Norway's environment, securing supply chains, and providing military advantage through expertise not otherwise available in the market.

Together with offset policies used in Norwegian defence acquisitions, the priority technology areas might lead to an increase in the amount of domestically manufactured components. However, the Norwegian approach is to encourage not only the domestic industry but also close cooperation with the international industry where such cooperation could lead to enhanced defence capabilities and security outcomes. This close cooperation is particularly evident in the development of weapon system technologies. Cooperation is also evident in relationships established to address platform integration, which Kongsberg clearly recognises as a critical element to the successful adoption of their weapons.

The foundation of the Norwegian approach has been innovation through close cooperation and commercialisation processes between the public S&T system and the R&D elements of defence companies. Kongsberg, in particular, has taken advantage of this innovation ecosystem, and *the value of such innovation is a key lesson for Australia.*

Kongsberg's success comes from a combination of innovating in certain areas and partnering with external entities that can deliver complementary areas of expertise, provide integration, and offer access to larger markets. If not for the expertise that grew from decades of developing missile systems, it is unlikely that Kongsberg would have developed world-class weapons; nor would it have achieved sustainable export success nor the close partnerships with U.S. and European defence companies, who otherwise would be considered competitors. Thus, *another lesson is that the best approach to establishing a sovereign weapons enterprise may be to focus on not only a sovereign (Australian) industry but also its critical relationships with global industry.*

In addition to garnering the advantages of mutually beneficial partnerships, *another lesson is that developing expertise and highly capable systems is not something that can happen quickly.* Kongsberg took ten years to develop the Penguin and even longer to develop more recent weapons. The company is now reaping the fruits of its early investments in missiles, with technological maturity and broad adoption of its missile technologies, some 60 years after work began on the Penguin.

One final lesson is that not only does development take time, so too may production of sophisticated weapons. There is no information in the public domain about Norway's options for surge production of the NSM and JSM, for example. However, there are indications that at least part of Norway's solution to the problem of surge production is to stockpile these missiles over time.[23]

---

[23] Aaron Mehta, 'Norway to Focus on Readiness, Stockpiling Munitions', *Defense News*, 31 January 2017.

CHAPTER SEVEN

Overall Lessons for Australia

We have identified relevant lessons for Australia from each of the case studies, although the different national circumstances means that no single case can apply as the 'solution' for Australia's Sovereign Guided Weapons Enterprise. Nonetheless, we draw seven overall lessons (or common themes from the case studies) for the establishment of such an enterprise. First and foremost, we recognise the complexity of creating such an enterprise and the need for such an enterprise to be bespoke to the Australian domestic and strategic context. Creating such an enterprise necessitates an agreed end state of what Australia is seeking to achieve, as the case studies revealed differing motivations. The case studies also highlighted that maturation of a weapons enterprise will take time and require support in terms of policies, innovation and education. The remaining overall lessons focus heavily on ensuring sustainable economic conditions for a sovereign enterprise, as well as prioritising partnerships and collaborations.

Table 7.1 offers a crosswalk between the 27 case study lessons and the seven overall lessons, showing which of the former apply to which of the latter. The left column shows the case study names. The right column shows 27 corresponding case study lessons. Each case study lesson appears beneath just one overall lesson. Nearly all the overall lessons were drawn from multiple case studies. The discussion following Table 7.1 describes each overall lesson in turn.

TABLE 7.1

Overall Lessons (Common Themes) Derived from Case Study Lessons

1: Define Desired Outcomes as Part of Developing a Bespoke Sovereign Enterprise	
Australia: ANSE	It is necessary to establish a clear vision for the enterprise—its end state.
Australia: ANSE	Metrics are needed to measure how well the enterprise is achieving its objectives.
2: Sovereignty Needs to Be Carefully Defined	
Japan: Guided Weapons Manufacturing	Foreign sales may be needed to maintain an affordable Sovereign Guided Weapons Enterprise.
Norway: Kongsberg Missile Programs (KMP)	The best approach to establishing a sovereign weapons enterprise may be to focus on not only a sovereign industry but also its critical relationships with global industry.
3: Complex Sovereign Enterprises Take Time and Effort to Build	
Canada: MSP	Mechanisms to confer confidence in munition requirements may not suffice for more-complex weapon systems involving more processes, supply chains and time.
Norway: KMP	Developing expertise and highly capable systems is not something that can happen quickly.

Table 7.1—Continued

3: Complex Sovereign Enterprises Take Time and Effort to Build	
Norway: KMP	Not only does development take time, so too may production of sophisticated weapons.
Australia: ANSE	A system to govern the many parties in a complex enterprise is necessary.
Australia: ANSE	Developing local supply chain management is challenging and expensive.
Australia: ANSE	An enterprise can involve multiple types of organisations—all with unique issues.
Australia: ANSE	Developing the design and engineering personnel team takes time.
4: Affordability and Sovereignty Need to Be Balanced	
Japan: Guided Weapons Manufacturing	Domestically developing and delivering a relatively small number of high-technology weapons will likely come with relatively high unit and sustainment costs.
UK: TCW	Developing custom guided munitions that are designed with cost-saving features, possibly with industry partnerships, can defray or eliminate the increased procurement costs of creating a sovereign program.
Australia: ANSE	Maintaining value-for-money is important for sustaining a commitment from the government and public.
Australia: MMI	The lowest-cost munitions for Australia may be from overseas vendors.
Australia: MMI	Purchasing munitions on the international market makes security of supply an issue.
5: Joint Development Offers Advantages and Limitations	
Japan: Guided Weapons Manufacturing	Joint development with allies and partner nations is attractive for a couple of reasons (wartime interoperability and shared learning).
UAE: Missiles and Weapons	An industrial defence park nurtured the cultivation of sovereign defence capabilities through interactions between domestic and international companies and the growth of local supply chains.
UK: TCW	Australia could consider partnering with companies that have different strengths to create new guided munition systems.
UK: TCW	Novel ways of partnering with industry or acquiring complex weapons may require new types of skills or additional resources within Defence.
Canada: Munitions Supply Program	Australia's integration with U.S. industry and access to it would need to not only match the levels already enjoyed by Canada but surpass them.
Norway: KMP	Innovation has come through close cooperation and commercialisation processes between the public S&T system and the R&D elements of defence companies.
6: Offsets Can Spur Growth Under Certain Circumstances	
UAE: Missiles and Weapons	Creating and continually adapting an offsets program enabled the defence sector growth and joint ventures that transferred critical technology.
Australia: ANSE	The AIC goals present challenges.
7: Industrial Capacity Needs to Be Right-Sized	
Australia: ANSE	Understand the long-term sustainability of the enterprise up front.
Australia: MMI	Think carefully about the design, development, engineering and production of munitions.
Australia: MMI	For a domestic munitions industry, using the full production capacity is the key to controlling costs.

Lesson 1: Define Desired Outcomes as Part of Developing a Bespoke Sovereign Enterprise

Maintaining control over the production and sustainment of essential defence capabilities is an existential concern for any state. Every capability system is unique, because each is subject to distinctive strategic, political, social, cultural and economic dynamics at the national, institutional and programmatic levels. As a result, despite apparent similarities in functional purposes and forms across states, each national defence acquisition system is bespoke to the nation it serves. In developing its own bespoke sovereign enterprise, Australia first needs to define and prioritise its desired outcomes. In doing so, Australia should carefully consider its strategic requirements across the range of guided weapon types and then prioritise its investments where they are most necessary or most relevant. This lesson may be particularly applicable to Australia's circumstances. Australia's strategic problems, situated at the fulcrum of the emerging Pacific Century, are unique in the international system. Australia's defence-related politics, its systems and institutions for defence acquisition, and its technological strengths and weaknesses are all distinct from those of even its closest allies. Australia's risks are also different, including the implications of its geography on the vulnerability and timeliness of external supply arrangements. The sovereignty of the weapons capability will therefore mean something different to Australia than it does to other nations.

Lesson 2: Sovereignty Needs to Be Carefully Defined

Sovereignty, in the case of guided weapons, is about removing the risks of losing access to, or control over, needed capabilities. This assured support to the military has been referred to as operational sovereignty. While Australia might need 'access to' or 'control over' the 'skills, technology' and other capabilities required for producing certain weapons, the capabilities do not necessarily need to be produced in Australia or by Australian companies to achieve operational sovereignty. Although the contribution of local industry to achieving sovereignty is important, it is also important to keep these two constructs—production and sovereignty—distinct from each other, as the analytic basis for decisionmaking in each regard is significantly different. Keys to operational sovereignty are not just the quality of the weapons technologies, but also the ability to replenish stock, which may be achieved not only through domestic production but also through inventory management and trusted supply chains. Additionally, Australia should consider which guided weapon types are most available on the global market and which will be the most difficult to acquire, either during a surge in demand or for replenishment. It might be more productive to focus on the weapon classes that are the most difficult to acquire.

Lesson 3: Complex Sovereign Enterprises Take Time and Effort to Build

Expertise to develop world-class weapons can take decades to develop, particularly when establishing a sustainable export business. Even with sufficient expertise, investments in complex weapon systems can take a decade or more to come to fruition, and that timeline seems to be only expanding as weapons become increasingly complex. It took Norway ten years to develop the Penguin and even longer for its more recent weapons. In addition to time, development of the enterprise requires substantial investment, as well as innovation and workforce support through external public entities involved in S&T and education. The over-all timelines and resources to develop sovereign weapons capabilities are further extended by the need either to develop sufficient surge capacity so that weapons can be produced quickly in time of need, or to have enough years of production to accumulate a sufficiently large stockpile. Beyond the length of time required is the ongoing effort required to govern all parties involved in a complex sovereign enterprise, including myriad different types of organisations inside and outside of government.

Lesson 4: Affordability and Sovereignty Need to Be Balanced

If Australia's domestic defence industry develops and manufactures a relatively small number of high-technology weapons for the ADF, the weapons will likely come with relatively high unit and sustainment costs. While this approach would enhance the sovereign defence capability and the security of supply chains, it would likely incur an above-market price premium. The relatively high costs would, in turn, limit the attractiveness of the weapons in the competitive international marketplace. This inability to effectively compete in the open market would further reinforce the small scale of weapon manufacturing. That is to say, although the cost of a domestic industry enterprise could be mitigated by its export potential, the export potential would in turn require large production volumes to reduce unit prices. From a sovereignty perspective, the likely higher unit costs of a domestically produced capability need to be balanced against the risks of having vulnerable weapon stocks and supply chains, and the resulting risks of mission failure.

Lesson 5: Joint Development Offers Advantages and Limitations

Joint development with allies and partner nations is attractive for a couple of reasons. Having common systems can aid interoperability and reduce localised wartime supply shortfalls. Each partner in a joint development team can also learn from the other partners—if information is shared freely. As evidenced in Norway, this mutual learning can include the cultivation of industry partner relationships that can complement technology development, reduce integration risk, and expand market demand. On the other hand, these partnerships are often stymied by concerns about proprietary information being unduly withheld, or conversely used inappropriately. Moreover, joint programs do not promote product differentiation, which can have negative strategic consequences and can also place Australian industry at a disadvantage. Strategically, while joint programs can increase the depth (or quantity) of a combined force's quiver, they do not increase the breadth (or variety) of that quiver. Economically, if Australia cannot produce a given system as cheaply as another locale, the Australian-produced munitions in a joint development effort will be disfavoured, given that an equivalent munition can be produced for lower cost elsewhere and still serve the joint effort.

Lesson 6: Offsets Can Spur Growth Under Certain Circumstances

A continually adapting offsets program can enable defence-sector growth and encourage joint ventures that result in the transfer of critical technology. Offset programs have been successfully used to spur the development and growth of domestic defence companies. Offsets have also facilitated joint ventures with international partners, and these joint ventures have provided critical technology transfers, including IP transfers, related to PGMs. Despite these notable successes, offset policies can impose additional costs on contracts and may not always achieve the aim of facilitating technology transfers and building sovereign capabilities. The AIC program already provides a framework whereby large contracts must contain plans for maximising the involvement of domestic companies; while not a formal offsets policy, the AIC program does seek to facilitate similar outcomes.

Lesson 7: Industrial Capacity Needs to Be Right-Sized

As Australia considers a Sovereign Guided Weapons Enterprise, its capacity should be consistent with domestic requirements. This capacity should include both peacetime inventory requirements and, if planned, an ability to surge for conflict. The production capacity should be driven by that required by Australia, and enough demand should be guaranteed to utilise the installed capacity. The alternative would be an orphaned production capacity that would unnecessarily increase unit costs and act as a deadweight within the guided weapon acquisition budget. Production costs could be reduced through export activities, but such activities would need to be permitted under technology controls, and there would need to be a realistic likely market. Thus, an attempt at right-sizing will require deliberate analysis, forecasts and decisions to produce a capability that is sufficient to meet Australian needs and adaptable to compete effectively in the international marketplace, but not too big as to increase fixed costs that cannot be recouped.

Abbreviations

ADF	Australian Defence Force
AIC	Australian Industry Content
ANSE	Australian Naval Shipbuilding Enterprise
ASM	Air-to-Ship Missile
ATLA	Acquisition, Technology and Logistics Agency
AUD	Australian dollar
AUSMIN	Australia–United States Ministerial Consultations
CAD	Canadian dollar
CAF	Canadian Armed Forces
CAMM	Common Anti-Air Modular Missile
COVID-19	coronavirus disease 2019
CWP	Complex Weapons Pipeline
Defence	Department of Defence (Australia)
DND	Department of National Defence
ESSM	Evolved Sea Sparrow Missile
FIC	fundamental inputs to capability
FMS	Foreign Military Sales
FOA	freedom of action
FY	fiscal year
GBP	Great British pounds
GDP	gross domestic product
IP	intellectual property
ITAR	International Traffic in Arms Regulations
Japan MOD	Japan Ministry of Defense
JSF	Joint Strike Fighter
JSM	Joint Strike Missile
KHI	Kawasaki Heavy Industries
KICs	Key Industrial Capabilities
MHI	Mitsubishi Heavy Industries
MMI	munitions manufacturing industry
MOD	Ministry of Defence (UK)
MSP	Munitions Supply Program
NAO	National Audit Office

NASAMS	National Advanced Surface to Air Missile System
NATO	North Atlantic Treaty Organization
NDPG	National Defense Program Guidelines
NORAD	North American Aerospace Defense Command
NSM	Naval Strike Missile
NTIB	National Technology and Industrial Base
ODCM	One Defence Capability Model
PGM	precision-guided missile
PIC	Priority Industry Capabilities
PMA	Portfolio Management Agreement
R&D	research and development
S&T	science and technology
SICAF	Sovereign Industry Capability Assessment Framework
SIPRI	Stockholm International Peace Research Institute
SPEAR	Selected Precision Effects at Range
TCW	Team Complex Weapons
UAE	United Arab Emirates
UOG	United Arab Emirates Offsets Group
USD	United States dollar

References

Acquisition.gov, 'Part 225—Foreign Acquisition', webpage, undated. As of 18 October 2021:
https://www.acquisition.gov/dfars/part-225-foreign-acquisition#DFARS-225.870

Acquisition, Technology and Logistics Agency, 'Missions of ATLA', webpage, undated. As of 10 August 2021:
https://www.mod.go.jp/atla/en/soubichou_gaiyou.html

Akimoto, Daisuke, 'Is Japan's Defense Industry in Decline?' *The Diplomat*, 1 October 2020. As of 10 August 2021:
https://thediplomat.com/2020/10/is-japans-defense-industry-in-decline/

ATLA—*See* Acquisition, Technology and Logistics Agency.

Axe, David, 'Japan Readies Its Ship-Smashing Super-Missile', *Forbes*, 16 July 2020. As of 12 September 2021:
https://www.forbes.com/sites/davidaxe/2020/07/16/japan-readies-its-ship-sinking-super-missile

Badr-El-Din, Amin, 'The Offsets Program in the United Arab Emirates', *Middle East Policy*, Vol. 5, No. 1, 1997, pp. 120–123.

Barry, Donald, and Duane Bratt, 'Defense Against Help: Explaining Canada-U.S. Security Relations', *American Review of Canadian Studies*, Vol. 38, No. 1, 2008, pp. 63–89.

Beazley, Kim, 'Innovative Boost to Exports from Defence Offsets', press release, 23 January 1987. As of 26 October 2021:
https://parlinfo.aph.gov.au/parlInfo/download/media/pressrel/HPR09026850/upload_binary/HPR09026850.pdf;fileType=application%2Fpdf#search=%22media/pressrel/HPR09026850%22

Berkok, Ugurhan, and Christopher Penney, *The Political Economy of the Munitions Supply Program*, Defence R&D Canada, Contract Report DRDC-RDDC-2014-C92, 2014.

Berthiaume, Lee, 'Canada Jumps Closer to Military-Spending Target Thanks to COVID-19's Economic Damage', *CCTV News*, 21 October 2020. As of 18 October 2021:
https://www.ctvnews.ca/politics/canada-jumps-closer-to-military-spending-target-thanks-to-covid-19-s-economic-damage-1.5154408

Binnie, Jeremy, 'IDEX 2021: Halcon to Provide C-RAM Missile for Rheinmetall for Air Defence', *Janes*, 23 February 2021.

Bitzinger, Richard, 'Military-Technical Innovation in Small States: The Cases of Israel and Singapore', *Journal of Strategic Studies*, June 2021.

Black, James, Stephen J. Flanagan, Gene Germanovich, Ruth Harris, David Ochmanek, Marina Favaro, Katerina Galai and Emily Ryen Gloinson, *Enhancing Deterrence and Defence on NATO's Northern Flank*, *Allied Perspectives on Strategic Options for Norway*, Santa Monica, Calif.: RAND Corporation, RR-4381-NMOD, 2020. As of 18 October 2021:
https://www.rand.org/pubs/research_reports/RR4381.html

Blinken, Antony J., Lloyd Austin, Marise Payne and Peter Dutton, 'Joint Press Availability', press release, 16 September 2021. As of 21 September 2021:
https://www.state.gov/secretary-antony-j-blinken-secretary-of-defense-lloyd-austin-australian-foreign-minister-marise-payne-and-australian-defence-minister-peter-dutton-at-a-joint-press-availability/

Bosbotinis, James, 'International Hypersonic Strike Weapons Projects Accelerate', *Aviation Week Intelligence Network*, 15 June 2020. As of 12 September 2021: https://aviationweek.com/defense-space/missile-defense-weapons/international-hypersonic -strike-weapons-projects-accelerate

Castellacci, Fulvio, and Arne Fevolden, 'Capable Companies or Changing Markets? Explaining the Export Performance of Firms in the Defence Industry', *Defence and Peace Economics*, Vol. 25, No. 6, 2014, pp. 549–575.

Castellacci, Fulvio, Arne Fevolden and Martin Lundmark, 'How Are Defence Companies Responding to EU Defence and Security Market Liberalization? A Comparative Study of Norway and Sweden', *Journal of European Public Policy*, Vol. 21, No. 8, 1218–1235.

Central Intelligence Agency, 'Canada', *The World Factbook*, last updated 14 September 2021a. As of 22 September 2021: https://www.cia.gov/the-world-factbook/countries/canada/

———, 'Norway: Military and Security Forces', *The World Factbook*, webpage, last updated 10 November 2021b. As of 22 September 2021: https://www.cia.gov/the-world-factbook/countries/norway/#military-and-security

Centre for Defence Industry Capability, 'Land Combat and Protected Vehicles Sovereign Industrial Capability Priority Plan Released', webpage, 3 May 2021.

Cheung, Tai Ming, 'A Conceptual Framework for Defence Innovation', *Journal of Strategic Studies*, June 2021.

Cheung, Tai Ming, Thomas G. Mahnken and Andrew L. Ross, 'Frameworks for Analyzing Chinese Defense and Military Innovation', in Tai Ming Cheung, ed., *Forging China's Military Might: A New Framework for Assessing Innovation*, Baltimore, Md.: Johns Hopkins University Press, 2014, pp. 15–46.

Combet, Greg, 'Launch of Australian Industry Capability Program', press release, 26 February 2008. As of 26 October 2021: https://parlinfo.aph.gov.au/parlInfo/search/display/display.w3p;query=Id%3A%22media %2Fpressrel%2FLIRP6%22;src1=sm1

Connolly, Amanda, and Kerri Breen, 'Canada "Not on Course" to Hit 2% Defence Spending Pledge: U.S. Official', *Global News*, 16 February 2020. As of 18 October 2021: https://globalnews.ca/news/6556192/canada-2-defence-spending-pledge

Cook, Cynthia R., Emma Westerman, Megan McKernan, Badreddine Ahtchi, Gordon T. Lee, Jenny Oberholtzer, Douglas Shontz and Jerry M. Sollinger, *Contestability Frameworks: An International Horizon Scan*, Santa Monica, Calif.: RAND Corporation, RR-1372-AUS, 2016. As of 26 October 2021: https://www.rand.org/pubs/research_reports/RR1372.html

Department of Defence, 'Defence White Paper', 2009. As of 26 October 2021: https://www.defence.gov.au/about/publications/2016-defence-white-paper

———, *2016 Defence Industry Policy Statement*, 2016. As of 26 October 2021: https://apo.org.au/sites/default/files/resource-files/2016-02/apo-nid93621.pdf

———, *Naval Shipbuilding Plan*, 2017. As of 26 October 2021: https://www.defence.gov.au/business-industry/naval-shipbuilding/plan

———, *2018 Defence Industrial Capability Plan*, 2018. As of 26 October 2021: https://www.defence.gov.au/business-industry/capability-plans/defence-industrial-capability-plan

———, *2020 Defence Strategic Update*, 2020a. As of 26 October 2021:
https://www.defence.gov.au/about/publications/2020-defence-strategic-update

———, *Defence Capability Manual*, 2020b. As of 26 October 2021:
https://defence.gov.au/publications/docs/Defence-Capability-Manual.pdf

———, 'Implementation and Industry Plans', webpage, 2020c. As of 15 October 2021:
https://www1.defence.gov.au/business-industry/programs/implementation-industry-plans

Department of National Defence, *Evaluation of Munitions Supply Program (MSP)*, Chief Review Services Document 1258-101-4, December 2007. As of 22 September 2021:
https://publications.gc.ca/collections/collection_2016/mdn-dnd/D58-164-2007-eng.pdf

DeVore, Marc, 'Armaments After Autonomy: Military Adaptation and the Drive for Domestic Defence Industries', *Journal of Strategic Studies*, May 2019, pp. 325–359.

Dibb, Paul, 'The Self-Reliant Defence of Australia: The History of an Idea', in Ron Huisken and Meredith Thatcher, eds., *History as Policy: Framing the Debate on the Future of Australia's Defence Policy*, Canberra: ANU Press, 2007, pp. 11–26.

DND—*See* Department of National Defence.

Defence—*See* Department of Defence.

Dominguez, Gabriel, 'Japan Awards Kongsberg Another Follow-On Contract for Joint Strike Missiles', *Jane's Defence Weekly*, 1 December 2020.

———, 'U.S. Department of State Approves Possible Sale of 73 SM-3 Block IIA Missile to Japan', *Jane's Defence Weekly*, 28 August 2019.

Donald, David, 'Denel and Tawazun Join Forces', *IHS: Africa Aerospace & Defence*, 21 September 2012.

Dortmans, Peter, Jennifer D. P. Moroney, Kate Cameron, Roger Lough, Emma Disley, Laurinda L. Rohn, Lucy Strang and Jonathan P. Wong, *Designing a Capability Development Framework for Home Affairs*, Santa Monica, Calif.: RAND Corporation, RR-2954-AUS, 2019. As of 22 October 2021:
https://www.rand.org/pubs/research_reports/RR2954.html

Dowse, Andrew, Tony Marceddo and Ian Martinus, 'Cyber Security and Sovereignty', *Australian Journal of Defence and Strategic Studies*, Vol. 3, No. 2, 2021.

Dunk, Graeme. 'Defence Industry Policy 2016—Well-Intentioned but Conflicted', *Security Challenges*, Vol. 12, No. 1, 2016, pp. 139–150.

———, 'The Decline of Trust in Australian Defence Industry', *Australian Defence Magazine*, 10 February 2020. As of 15 October 2021:
https://www.australiandefence.com.au/news/the-decline-of-trust-in-australian-defence-industry

EDGE Group, 'Al Jasoor Joins Forces with Raytheon Emirates, EARTH to Integrate High Energy Laser Systems onto Rabdan Vehicles', webpage, undated. As of 22 September 2021:
https://edgegroup.ae/news/585

Fackler, Martin, 'Japan Ends Decades-Long Ban on Export of Weapons', *New York Times*, 2 April 2014.

Fevolden, Arne, and Kari Tvetbråten, 'Defence Industrial Policy—A Sound Security Strategy or an Economic Fallacy?' *Defence Studies*, 2016, Vol. 16, No. 2, pp. 176–192.

Fleurant, Aude, Alexandra Kuimova, Diego Lopes Da Silva, Nan Tian, Pieter D. Wezeman, and Siemon T. Wezeman, 'The SIPRI Top 100 Arms-Producing and Military Services Companies, 2018', SIPRI Fact Sheet, Solna, Sweden: Stockholm International Peace Research Institute, December 2019.

Freeman, Alan, 'Minister: Canada Will Build Up Its Military as the U.S. Pulls Back from World Stage', *Washington Post*, 6 June 2017. As of 22 September 2021:
https://www.washingtonpost.com/world/the_americas/minister-canada-will-build-up-its-military-as-the-us-pulls-back-from-world-stage/2017/06/06/4c841a22-4ad8-11e7-987c-42ab5745db2e_story.html

Gady, Franz-Stefan, 'U.S. State Department Greenlights $3.3 Billion Missile Sale to Japan', *The Diplomat*, 12 September 2019. As of 12 September 2021:
https://thediplomat.com/2019/09/us-state-department-greenlights-3-3-billion-missile-sale-to-japan/

Gat, Azar, *A History of Military Thought*, Oxford: Oxford University Press, 2001.

Gaub, Florence, and Zoe Stanley-Lockman, *Defence Industries in Arab State: Players and Strategies*, European Union Institute for Security Policy Studies, Chaillot Paper No. 141, March 2017. As of 26 October 2021:
https://www.iss.europa.eu/sites/default/files/EUISSFiles/CP_141_Arab_Defence.pdf

Gholz, Eugene, and Harvey Sapolsky, 'The Defense Innovation Machine: Why the U.S. Will Remain on the Cutting Edge', *Journal of Strategic Studies*, June 2021.

Government of Canada, *Statement on Canada's Arctic Foreign Policy*, 2010. As of 14 October 2021:
https://www.international.gc.ca/world-monde/assets/pdfs/canada_arctic_foreign_policy-eng.pdf

———, *Canada First: Leveraging Defence Procurement Through Key Industrial Capabilities*, February 2013. As of 14 October 2021:
https://www.tpsgc-pwgsc.gc.ca/app-acq/amd-dp/documents/eam-lmp-eng.pdf

———, *Strong, Secure, Engaged: Canada's Defence Policy*, 2017. As of 18 October 2021:
http://dgpaapp.forces.gc.ca/en/canada-defence-policy/docs/canada-defence-policy-report.pdf

———, 'Defence Capability Blueprint', webpage, last updated 9 January 2020. As of 18 October 2021:
http://dgpaapp.forces.gc.ca/en/defence-capabilities-blueprint/index.asp

———, 'Munitions Supply Program', webpage, last updated 28 April 2021. As of 22 September 2021:
https://www.tpsgc-pwgsc.gc.ca/app-acq/amd-dp/munitions-eng.html

Government of Norway, 'Norwegian Exports of Defence-Related Products in 2020', webpage, 11 June 2021. As of 22 September 2021:
https://www.regjeringen.no/en/aktuelt/exports_2020/id2860735/

Government of South Australia, 'Submission 66 to the Joint Standing Committee on Foreign Affairs, Defence and Trade Inquiry into the Implications of the COVID-19 Pandemic', 2020. As of 18 October 2021:
https://www.aph.gov.au/Parliamentary_Business/Committees/Joint/Foreign_Affairs_Defence_and_Trade/FADTandglobalpandemic/Submissions

Greenwalt, William, *Leveraging the National Technology Industrial Base to Address Great Power Competition: The Imperative to Integrate Industrial Capabilities of Close Allies*, Atlantic Council Scowcroft Center for Strategy and Security, April 2019. As of 22 September 2021:
https://www.atlanticcouncil.org/wp-content/uploads/2019/04/Leveraging_the_National_Technology_Industrial_Base_to_Address_Great-Power_Competition.pdf

Grevatt, Jon, 'Japan Pays a Premium for Locally Built F-35s', *Jane's Defence Weekly*, 14 March 2018.

———, 'Japanese MoD Reviews Procurement Priorities', *Jane's Defence Industry*, 7 July 2020.

———, 'Japan Commits to Local F-35 Production', *Jane's Defence Weekly*, 30 July 2020.

———, 'Japan Looks to Introduce Finance System for Defence Experts', *Jane's Defence Industry*, 18 May 2021.

Hatlebakk Hove, Kjetil, *Defence Industrial Policy in Norway: Drivers and Influence*, Armament Industry Europe Research Group, Paper 25, February 2018. As of 18 October 2021:
https://www.iris-france.org/wp-content/uploads/2018/02/Ares-25-Policy-Paper-f%C3%A9vrier-2018.pdf

Helou, Agnes, 'Saab Inks Deal with Tawazun to Bolster Sensor Technology Research in UAE', *Defense News*, 22 February 2021. As of 25 August 2021:
https://www.defensenews.com/digital-show-dailies/idex/2021/02/22/saab-inks-deal-with-tawazun-to-bolster-sensor-technology-research-in-uae/

Herman, Arthur, *The Awakening Giant: Risk and Opportunities for Japan's New Defense Export Policy*, Washington, D.C.: Hudson Institute, December 2016.

Hix, William M., Bruce Held and Ellen M. Pint, *Lessons from the North: Canada's Privatization of Military Ammunition Production*, Santa Monica, Calif: RAND Corporation, MG-169-OSD, 2004. As of 22 September 2021:
https://www.rand.org/pubs/monographs/MG169.html

House of Commons Defence Committee, *Defence Equipment 2008: Tenth Report of Session 2007–08*, 27 March 2008. As of 26 August 2021:
https://publications.parliament.uk/pa/cm200708/cmselect/cmdfence/295/295.pdf

House of Commons Select Committee on Defence, *Sixth Report*, 19 December 2006. As of 26 August 2021:
https://publications.parliament.uk/pa/cm200607/cmselect/cmdfence/177/17705.htm

Howell, Sabrina T., Jason Rathje, John Van Reenen and Jun Wong, 'Opening Up Military Innovation: Causal Effects of "Bottom-Up" Reforms to U.S. Defense Research', IZA—Institute of Labor Economics Discussion Paper No. 14297, April 2021. As of 26 October 2021:
https://docs.iza.org/dp14297.pdf

Hoyle, Craig, 'Farnborough: MBDA to Head UK Arms Team', *Flight Global*, 24 July 2006. As of 25 August 2021:
https://www.flightglobal.com/farnborough-mbda-to-head-uk-arms-team/68707.article

Hunter, Andrew, Kristina Obecny and Gregory Sanders, *U.S.-Canadian Defense Industrial Cooperation*, Center for Strategic and International Studies, June 2017. As of 18 October 2021:
https://www.csis.org/analysis/us-canadian-defense-industrial-cooperation

Innovation, Science and Economic Development Canada, *State of Canada's Defence Industry*, 2018. As of 18 October 2021:
https://www.ic.gc.ca/eic/site/ad-ad.nsf/vwapj/StateCanadaDefence2018Report.pdf/$file/StateCanadaDefence2018Report.pdf

Insinna, Valerie, 'State Department Clears $113M Sale of AMRAAM Missiles to Japan', *DefenseNews*, 5 October 2017. As of 12 September 2021:
https://www.defensenews.com/global/asia-pacific/2017/10/04/state-department-clears-113m-sale-of-amraam-missiles-to-japan/

International Trade Administration, 'Norway—Country Commercial Guide', webpage, last updated 10 October 2021. As of 26 October 2021:
https://www.trade.gov/country-commercial-guides/norway-defense-and-aerospace-technologies

Jain, Purnendra, 'Japan's Weapon-Export Industry Takes Its First Steps', *East Asia Forum*, 30 October 2020. As of 18 August 2021:
https://www.eastasiaforum.org/2020/10/30/japans-weapon-export-industry-takes-its-first-steps/

Janes Weapons, 'Naval Strike Missile (NSM)', webpage, 17 February 2020. As of 18 October 2021: https://customer.janes.com/Janes/Display/JNWS0911-JNW_

———, 'NASAMS', webpage, 19 November 2020. As of 18 October 2021: https://customer.janes.com/Janes/Display/JLAD0223-JAAD

———, 'Joint Strike Missile (JSM)', webpage, 6 July 2021. As of 18 October 2021: https://customer.janes.com/Janes/Display/JALW3691-JALW

Janes World Air Forces, 'Canada—Air Force', webpage, last updated 13 May 2021a. As of 22 September 2021: https://customer.janes.com/WorldAirForces/Display/jwafa044-JWAF#Core%20assets%20and%20procurement%20initiatives

———, 'Norway—Air Force', webpage, last updated 17 August 2021b. As of 26 October 2021: https://customer.janes.com/WorldAirForces/Display/jwafa198-JWAF

Janes World Armies, 'Canada—Army', webpage, last updated 7 May 2021. As of 21 September 2021: https://customer.janes.com/WorldArmies/Display/JWARA128-JWAR#Summary

Janes World Navies, 'Norway—Navy' webpage, last updated 14 April 2021a. As of 18 October 2021: https://customer.janes.com/WorldNavies/Display/JWNA0115-JWNA

———, 'Canada—Navy', webpage, last updated 13 May 2021b. As of 22 September 2021: https://customer.janes.com/WorldNavies/Display/JWNA0029-JWNA#Summary

Japan Ministry of Defense, *Introduction to the Equipment of the Japan Self-Defence Forces: A Reference Guide to the Defense Industrial Base of Japan (2nd Edition)*, Acquisition, Technology, and Logistics Agency, undated. As of 12 September 2021: https://www.mod.go.jp/atla/soubiseisaku/soubiseisakugijutu/introduction2020_en.pdf

———, *National Defense Program Guidelines for FY 2019 and Beyond (Provisional Translation)*, 18 December 2018. As of 26 October 2021: https://warp.da.ndl.go.jp/info:ndljp/pid/11591426/www.mod.go.jp/j/approach/agenda/guideline/2019/pdf/20181218_e.pdf

———, *Defense of Japan*, 2021. As of 26 October 2021: https://www.mod.go.jp/en/publ/w_paper/wp2021/DOJ2021_Digest_EN.pdf

Japan Ministry of Foreign Affairs, 'Three Principles on Transfer of Defense Equipment and Technology', webpage, 6 April 2016. As of 9 August 2021: https://www.mofa.go.jp/fp/nsp/page1we_000083.html

Japan MOD—*See* Japan Ministry of Defense.

Japan National Security Council, 'Implementation Guidelines for the Three Principles on Transfer of Defense Equipment and Technology', 1 April 2014. As of 1 December 2021: https://www.mofa.go.jp/files/000121050.pdf

Joint Standing Committee on Foreign Affairs, Defence and Trade, *Principles and Practice—Australian Defence Industry and Exports: Inquiry of the Defence Sub-Committee*, Parliament of the Commonwealth of Australia, 2015. As of 18 October 2021: https://www.aph.gov.au/Parliamentary_Business/Committees/Joint/Foreign_Affairs_Defence_and_Trade/Defence_Industry_Exports/Report

Kongsberg, '200 Years of Determination', webpage, undated a. As of 18 October 2021: https://www.kongsberg.com/kda/Who-we-are/200-years-of-excellence/

———, 'Key Figures 2020', webpage, undated b. As of 18 October 2021: https://www.annual-report.kongsberg.com/year-2020/key-figures-2020/

Krasner, Stephen, 'Rethinking the Sovereign State Model', *Review of International Studies*, Vol. 27, 2001, pp. 15–42.

Markowski, Stefan, and Peter Hall, 'Defence Procurement and Industry Development: Some Lessons from Australia', in Ugurhan G. Berkok, ed., *Studies in Defence Procurement*, Kingston, Ont.: Queens, 2006, pp. 9–73.

Marshall, Shana, 'The New Politics of Patronage: The Arms Trade and Clientalism in the Arab World', Brandeis Crown Center Working Paper 4, 2012.

MBDA Missile Systems, 'MBDA & UK Mod, Long Term Partnering for Complex Weapons', press release, 29 March 2010. As of 26 August 2021:
https://www.mbda-systems.com/press-releases/mbda-uk-mod-partnering-for-complex-weapons/

———, 'Achieving Benefits', webpage, 2021a. As of 26 August 2021:
https://www.mbda-systems.com/about-us/mission-strategy/team-complex-weapons/achieving
-benefits/

———, 'Team Complex Weapons', webpage, 2021b. As of 25 August 2021:
https://www.mbda-systems.com/about-us/mission-strategy/team-complex-weapons/

Mehta, Aaron, 'Norway to Focus on Readiness, Stockpiling Munitions', *Defense News*, 1 February 2017. As of 22 September 2021:
https://www.defensenews.com/air/2017/01/31/norway-to-focus-on-readiness-stockpiling-munitions/

Mezher, Chyrine, 'UAE's First Air Defense Missile to Be Used on German Oerlikon Skynex', *Breaking Defence*, 24 February 2021. As of 26 October 2021:
https://breakingdefense.com/2021/02/uaes-first-air-defense-missile-to-be-used-on-german
-oerlikon-skynex/

Miki, Rieko, 'The Price of Peace: Why Japan Scrapped a $4.2bn U.S. Missile System', *NikkeiAsia*, 5 August 2020. As of 10 August 2021:
https://asia.nikkei.com/Spotlight/The-Big-Story/The-price-of-peace-Why-Japan-scrapped-a
-4.2bn-US-missile-system

Ministry of Defence, *Defence Industrial Strategy*, Defence White Paper, December 2005. As of 24 August 2021:
https://assets.publishing.service.gov.uk/government/uploads/system/uploads/attachment_data/
file/272203/6697.pdf

———, 'MoD Launches a New Approach to Acquiring Complex Weapons', press release, 25 July 2008. As of August 26, 2021:
https://www.wired-gov.net/wg/wg-news-1.nsf/0/38299592A2CAF64D80257487003CF6D5?Open
Document

———, *National Security Through Technology: Technology Equipment, and Support for UK Defence and Security*. As of 26 August 2021:
https://assets.publishing.service.gov.uk/government/uploads/system/uploads/attachment_data/
file/27390/cm8278.pdf

———, *The Defence Equipment Plan 2019: Financial Summary*, 27 February 2020. As of 26 August 2021:
https://assets.publishing.service.gov.uk/government/uploads/system/uploads/attachment_data/
file/930062/20201028_EP19_v2_Official.pdf

———, '£550 F-35 Missile Contract Signed', webpage, 6 January 2021a. As of 26 August 2021:
https://des.mod.uk/500m-f35-missile-contract-signed

——, 'The Competition Act 1998, Public Policy Exclusion Order 2007 No. 1896: Complex Weapons', webpage, 11 January 2021b. As of 26 August 2021:
https://www.legislation.gov.uk/ukia/2011/434/pdfs/ukia_20110434_en.pdf

MOD—*See* Ministry of Defence.

Morgenthau, Hans J., *Politics Among Nations: The Struggle for Power and Peace*, New York: Knopf, 1948.

NAO—*See* National Audit Office.

National Audit Office, *Ministry of Defence—The Major Projects Report 2012*, HC 684, 2012–2013, 10 January 2013. As of 26 August 2021:
https://www.nao.org.uk/report/ministry-of-defence-the-major-projects-report-2012/

——, *The Major Projects Report 2013*, Ministry of Defence Report, HC 817-I, Session 2013–2014, 13 February 2014. As of 26 August 2021:
https://www.nao.org.uk/wp-content/uploads/2015/02/The-Major-Projects-Report-2013.pdf

——, *Major Projects Report 2015 and the Equipment Plan 2015 to 2025*, Ministry of Defence Report, HC 488-I, Session 2015–2016, 22 October 2015.

——, *The Equipment Plan 2016 to 2026*, Ministry of Defence Report, HC 914, Session 2016–2017, 27 January 2017. As of 26 August 2021:
https://www.nao.org.uk/wp-content/uploads/2017/01/The-Equipment-Plan-2016-2026.pdf

——, *The Equipment Plan 2019 to 2029*, Ministry of Defence Report, HC 111, Session 2019–2020, 27 February 2020. As of 26 August 2021:
https://www.nao.org.uk/wp-content/uploads/2020/02/The-Equipment-Plan-2019-to-2029.pdf

NATO, 'Defence Expenditure of NATO Countries (2014–2021)', press release, 11 June 2021a. As of 21 October 2021:
https://www.nato.int/nato_static_fl2014/assets/pdf/2021/6/pdf/210611-pr-2021-094-en.pdf

——, 'Funding NATO', webpage, last updated 13 August 2021b. As of 22 September 2021:
https://www.nato.int/cps/en/natohq/topics_67655.htm

Norwegian Ministry of Defence, *The Defence of Norway: Capability and Readiness, Long Term Defence Plan 2020*, 17 April 2020. As of 22 September 2021:
https://www.regjeringen.no/contentassets/3a2d2a3cfb694aa3ab4c6cb5649448d4/long-term
-defence-plan-norway-2020---english-summary.pdf

——, *Cooperation for Security: National Defence Industrial Strategy: A Technologically Advanced Defence for the Future*, Meld. St. 17 (2020–2021), 12 March 2021. As of 22 September 2021:
https://www.regjeringen.no/contentassets/5f29db6ef1b34054a025ffddb7073b31/en-gb/pdfs/
stm202020210017000engpdfs.pdf

Norwegian Ministry of Trade, Industry and Fisheries, *The State's Direct Ownership of Companies: Sustainable Value Creation*, Meld. St. 8 (2019–2020), 22 November 2019. As of 18 October 2021:
https://www.regjeringen.no/contentassets/44ee372146f44a3eb70fc0872a5e395c/en-gb/pdfs/
stm201920200008000engpdfs.pdf

Nose, Miki, 'Politics: Japan Defense Spending Isn't Bound by 1% GDP Cap, Suga Says', *Nikkei Asia*, 13 August 2021. As of 18 August 2021:
https://asia.nikkei.com/Politics/Japan-defense-spending-isn-t-bound-by-1-GDP-cap-Suga-says

Office of the Secretary of Defense, *Cost Assessment and Program Evaluation*, September 2020. As of 21 September 2021:
https://www.cape.osd.mil/files/OS_Guide_Sept_2020.pdf

Pearl, Daniel, 'Offset Requirements of Defense Deals Often Have Little to Do with Purchaser', *Wall Street Journal*, 20 April 2000.

Petersson, Magnus, 'Small States and Autonomous Systems—The Scandinavian Case', *Journal of Strategic Studies*, December 2020, pp. 594–612.

Porter, Michael E., 'Clusters and the New Economics of Competition', *Harvard Business Review*, November–December 1998.

Price, Melissa, 'The Multifaceted Benefits of the Sovereign Guided Missiles Project', *Defence Connect*, 10 August 2021. As of 27 August 2021:
https://www.defenceconnect.com.au/blog/8542-the-multifaceted-benefits-of-the-sovereign-guided-missiles-project

Prime Minister, Minister for Defence, Minister for Defence Industry, Minister for Industry and Minister for Industry Science and Technology, 'Sovereign Guided Weapons Manufacturing', press release, 31 March 2021. As of 21 September 2021:
https://www.pm.gov.au/media/sovereign-guided-weapons-manufacturing

Purnell, Leigh, and Mark Thomson, *How Much Information is Enough?*, Australian Strategic Policy Institute, December 2009. As of 20 September 2021:
https://s3-ap-southeast-2.amazonaws.com/ad-aspi/import/ASPI_DCP_Review_Report_web.pdf?VersionId=bUl9rdwj0rXcq94yoXsbUC6mX6bQoleF

Roberts, Corry, 'Australia's Sovereign Guided Weapons Heritage', *Thales*, 19 July 2021. As of 18 October 2021:
https://www.thalesgroup.com/en/australia/news/australias-sovereign-guided-weapons-heritage

Sakaki, Alexandra, and Sebastian Maslow, 'Japan's New Arms Export Policies: Strategic Aspirations and Domestic Constraints', *Australian Journal of International Affairs*, Vol. 74, No. 6, 2020, pp. 649–669.

Schwartz, Paul, *The Changing Nature and Implications of Russian Military Transfers to China*, Center for Strategic International Studies, June 2021. As of 26 August 2021:
https://csis-website-prod.s3.amazonaws.com/s3fs-public/publication/210621_Schwartz_Russian_Military_Transfers.pdf?47lttXU2w57d.CobDxg1b1nGmtA1tUcU

SIPRI—*See* Stockholm International Peace Research Institute.

Stanton, Blake, 'Reducing Weapon Costs Through the Team Complex Weapons Portfolio Agreement', Atkins, 6 February 2018. As of 22 September 2021:
https://www.aerosociety.com/media/8057/raes_reducing-weapon-costs_tcw_6feb.pdf

Stockholm International Peace Research Institute, 'Global Arms Industry: Sales by the Top 25 Companies up 8.5 Per Cent; Big Players Active in Global South', press release, 7 December 2020. As of 19 August 2020:
https://www.sipri.org/media/press-release/2020/global-arms-industry-sales-top-25-companies-85-cent-big-players-active-global-south

Strassler, Robert B., *The Landmark Thucydides: A Comprehensive Guide to the Peloponnesian War*, New York: Free Press, 1998.

Takahashi, Kosuke, 'Update: Tokyo Approves Plan to Develop Type 12 Missile into Stand-Off Weapon', *Jane's Defence Weekly*, 18 December 2020a.

———, 'Japan Moves Ahead with JNAAM Co-Development', *Jane's Missiles & Rockets*, 22 December 2020b.

———, 'Japan to Bring Mass Production of New ASM-3A Supersonic Anti-Ship Missile', *Jane's Defence Weekly*, 4 January 2021.

Thomas-Noone, Brendan, *Ebbing Opportunity: Australian and the U.S. National Technology and Industrial Base*, U.S. Studies Centre, November 2019. As of 22 September 2021:
https://www.ussc.edu.au/analysis/australia-and-the-us-national-technology-and-industrial-base

Thompson, Marcus, 'Information Warfare—a New Age?' speech delivered at the Military Communications and Informations Systems Conference, Canberra, Australia, 15 November 2018. As of 15 October 2021:
https://defence.gov.au/JCG/docs/MILCIS2018-HIW-Transcript.pdf

Tilly, Charles, *Coercion, Capital, and European States, AD 990–1992*, London: Blackwell, 1992.

Waltz, Kenneth N., *Theory of International Politics*, New York: McGraw-Hill, 1979.

World Bank, 'Military Expenditure (% of GDP)—Canada', webpage, undated a. As of 22 September 2021:
https://data.worldbank.org/indicator/MS.MIL.XPND.GD.ZS?locations=CA

———, 'Military Expenditure (% of GDP)—Norway', webpage, undated b. As of 18 October 2021:
https://data.worldbank.org/indicator/MS.MIL.XPND.GD.ZS?locations=NO

Yeo, Mike, 'Japan to Cease In-Country Assembly of F-35 Jets', *DefenseNews*, 18 January 2019. As of 12 September 2021:
https://www.defensenews.com/industry/2019/01/17/japan-to-cease-in-country-assembly-of-f-35-jets/